D1276487

WAR AND THE
IVORY TOWER

WAR AND THE IVORY TOWER

Algeria and Vietnam

DAVID L. SCHALK

New York   Oxford
OXFORD UNIVERSITY PRESS
1991

Oxford University Press

Oxford   New York   Toronto
Delhi   Bombay   Calcutta   Madras   Karachi
Petaling Jaya   Singapore   Hong Kong   Tokyo
Nairobi   Dar es Salaam   Cape Town
Melbourne   Auckland

and associated companies in
Berlin   Ibadan

Copyright © 1991 by Oxford University Press, Inc.

Published by Oxford University Press, Inc.,
200 Madison Avenue, New York, New York 10016

Oxford is a registered trademark of Oxford University Press

All rights reserved. No part of this publication may be reproduced,
stored in a retrieval system, or transmitted, in any form or by any means,
electronic, mechanical, photocopying, recording, or otherwise,
without the prior permission of Oxford University Press.

Library of Congress Cataloging-in-Publication Data
Schalk, David L.
War and the ivory tower : Algeria and Vietnam / David L. Schalk.
p. cm.
Includes bibliographical references and index.
ISBN 0-19-506807-6
1. Algeria—History—Revolution, 1954–1962—Protest movements—France.
2. Vietnamese Conflict, 1961–1975—Protest movements—United States.
3. Peace movements—France.
4. Peace movements—United States.   I. Title.
DT295.S354   1991
959.704'3373—dc20
90-21086

2 4 6 8 9 7 5 3 1

Printed in the United States of America
on acid-free paper

DT
295
.S354

On October 2, 1984, Jerry Serino, president of the Mid-Hudson Chapter of the Vietnam Veterans of America, stood over his father's grave in my home town of Poughkeepsie, New York, and put a bullet through his chest. The note he left in his shirt pocket said simply: "America, don't forget Vietnam. Future generations deserve to know." The work that follows was conceived as a fragment of a response to Mr. Serino's last request.

If through comparative historical analysis I am also able to reach French veterans of the Algerian War and help reduce their "silence and shame," which has persisted for more than a quarter-century, I shall be doubly gratified.

Acknowledgments

After I had done some background reading and realized the extraordinary political and military linkages between the Algerian and Vietnam wars—both undeclared and seeming to drag on endlessly, separated by almost exactly a decade, I devoted considerable attention to developing a network of correspondents who had expertise in either Algeria or Vietnam. I have received much enthusiastic support and many helpful bibliographical suggestions from researchers on both sides of the Atlantic, and I owe them a debt of gratitude for superb work that they have done in their respective provinces. I have, however, never been able to locate anyone involved in a comparative project similar to mine. Hence this book is the result of the research and reflection of a solitary scholar who takes full responsibility for all the arguments and conclusions presented.

Six individuals do deserve special thanks for their encouragement and inspiration: Louis Gardel and Arno Mayer just when I was developing a conceptual framework; Carl Schorske when I was completing my research; and Ali Haroun and David W. Levy toward the end of my writing. Before anyone else did, Nancy Lane of Oxford University Press understood both the potential and the difficulties inherent in this kind of study. From the beginning she has been a guide and mentor of unsurpassed skill and knowledge.

Part of Chapter 3 was published in *The Tocqueville Review*, vol. 8 (1986–87), along with an extremely valuable commentary by Jean-Marie Domenach. I am indebted to Jesse Pitts, editor of *The Tocqueville Review*, and to M. Domenach for their interest and wise counsel. Some of the key arguments in this book were summarized in a paper read in French at the Institut d'histoire du temps présent in April 1988 and published by the Institut in cahier no. 10 in November 1988. I am very grateful to Jean-Pierre Rioux and his colleagues at the Institut for their hospitality and their generosity in sharing with me their own findings and conclusions on question of intellectual engagement in the latter half of the twentieth century.

D. L. S.

Poughkeepsie, N. Y.
August 1990

Contents

WAR AND THE
IVORY TOWER

Introduction

"*Intellectual*, noun, masculine gender, a social and cultural category born in Paris at the moment of the Dreyfus affair, dead in Paris at the end of the twentieth century; apparently was not able to survive the decline of belief in Universals."[1] Such is the definition from a dictionary of the year 2000, as imagined by the "new philosopher" Bernard-Henri Lévy in an intriguing little book published in 1987, *Éloge des Intellectuels*. Lévy, somewhat of a media star who thrives on controversy, told the *New York Times* that "France is a country where the glory of literature has always been linked by a concrete, massive engagement in the affairs of the century."[2] Now, Lévy believes, the engagement has ended, and intellectuals are no longer hated and assaulted as they were in earlier times, such as the period of the Algerian War. The crisis that intellectuals are going through is a quiet collapse, a *débâcle*, intimately linked with their withdrawal from the public stage and their return to the ivory tower. It will probably lead to their disappearance, Lévy argues. Surviving intellectuals suffer from a sense of unreality; in the country of Voltaire and Zola, business people, singers, and actors are now consecrated as *maîtres à penser*.[3] One of the goals of this book is to shed some light on the question of whether intellectuals as a species are indeed about to become extinct, and not only in France.

Though I approach this question from a different political perspective, I share many of Lévy's concerns and agree with

3

him that the presence of intellectuals in a modern society is a "key to democracy."[4] Hence I am not especially happy to find ideas that I articulated a decade ago entering the realm of popular culture and being sanctified by the *New York Times*. I did argue in *The Spectrum of Political Engagement*, which appeared in 1979, that there seems to be a close, though complex, link between the intellectual, as opposed to the "mandarin" or "specialist," and engagement and that the one may be impossible without the other. Writing at a time when disengagement was already widespread among the intellectual classes in Europe and North America, I thought it possible that the appearance of intellectuals (and thus the manifestation of engagement, of critical dissent) was "a passing phenomenon, born on the streets of Paris in 1898 and dying there exactly seventy years later."[5]

My training has been as a European (especially French) intellectual historian, and *The Spectrum of Political Engagement* dealt with the political involvement of French intellectuals in the period between 1930 and 1945. One reviewer suggested that at a deeper level my book addressed the question of the "power of intellectuals" and thought that the "specter of Vietnam" lay behind every page.[6] Authors can often learn from reviewers, and if *Spectrum* interested a wider range of readers than the usual scholarly monograph would, the reason may be the presence of a "specter" of which I was, at most, minimally aware.

When I completed *Spectrum*, my initial plan was to write a sequel concentrating on France, examining the passionate debate over engagement and the responsibility of intellectuals that raged there between 1945 and 1960. I intended to choose as a centerpiece the bitter quarrel between France's two most famous intellectuals, Albert Camus and Jean-Paul Sartre, a feud that had simmered since 1945 and became painfully public in 1951 when Camus's *L'Homme révolté* (The rebel) appeared. It lasted until Camus's untimely death in an automobile accident in 1960.[7]

Gradually I came to the conclusion that a more original and fruitful project would involve skipping a decade and beginning my study of engagement with the outbreak of the undeclared war in Algeria in November 1954. I had lived in France during 1961 and 1962 when the Algerian conflict reached its paroxysm of violence and had vivid memories of the acrid smell of plastic explosives wafting through the streets of Paris and of the omnipresence of heavily armed security troops who periodically pointed semiautomatic weapons at me and demanded that I open my briefcase to see if I were carrying a bomb. I also had personal contacts with a number of *pieds-noirs*, ethnic Frenchmen and women born in Algeria.

One *pied-noir*, the distinguished writer Louis Gardel, author of three remarkable novels set in Algeria,[8] told me of his student days at the Lycée Louis le Grand in Paris in 1956 and how antiwar activists asked him to sign a petition protesting the use of torture by French troops in Algeria. He agreed to do so if he could add a sentence condemning what were, from his perspective, the equally brutal murders of civilians and captured French soldiers committed by the irregulars of the FLN (Front de libération nationale). Regrettably, in an example of the kind of tunnel vision that bothered many sincere opponents of America's war in Vietnam, his signature was refused.

At this precise moment I conceived the extravagant, perhaps foolhardy, notion of drawing the parallels between Algeria and Vietnam, focusing on the intellectual antiwar engagement in the two cases. I had been a young academic during the chaotic and often agonizing years of our Vietnam War and had firsthand knowledge of the intellectual opposition to that war. I suggested to Louis Gardel that he write the Algerian part of a coauthored book. But he replied that he could not, for his emotions were too deeply involved, and so with his encouragement I embarked on the project alone, assuming that as a trained historian I could at least subdue, if not eliminate, my own powerful feelings. Along with the great majority of my colleagues, I have shared "that noble dream" of historical objectivity, attempting to ignore or circumvent the warnings issued by Charles Beard in a famous 1935 article by the same title. Peter Novick has shown in a very important book, brilliantly yet deftly deflating the pretensions of the American historical profession, that the dream, which we imported from Germany in the nineteenth century, has become somewhat of a nightmare.[9]

There have been several other pitfalls in my project, among them the very decision to write comparative history, which is a difficult genre and one not often practiced. In his presidential address to the American Historical Association in December 1986, Carl Degler made an eloquent appeal to his colleagues to try it, even though "when one advocates...that the United States be compared historically with other societies, the specter of American exceptionalism [a different specter!] inevitably floats before us." Degler urged us to reconsider the notion that our country exists and has existed completely separate from European patterns of development, citing a wise warning made more than a half-century ago by the great French historian Marc Bloch, that the identification of differences is a primary reason for historical comparison.[10]

A third difficulty I faced in devising a plan for my research was whether or not to employ the exciting, often productive, and popular route of oral history. Many engaged intellectuals from both countries are still alive and could be interviewed. Some important work has already been done in France, where two groups of antiwar intellectuals, one from the independent Left and the other present and former members of the French Communist party, have been carefully interviewed and the findings published. The "decantation operated by memory," as Rémy Reiffel, one of the authors, elegantly puts it, and the "justifications intellectuals gave in 1987–1988 for their past engagements" are fascinating.[11]

I concluded fairly quickly, however, that the oral history

approach would not be appropriate to the project I envisioned, which had to involve a close examination of primary sources. Recollections of events now fairly distant filtered through the memories of participants have an undeniable appeal for historians and psychologists and a powerful general human interest. But they reflect the passage of years and fresh hopes and disillusions, and even if they do not deliberately distort, they cast a different light on thoughts and actions as they occurred in their historical contexts. To illustrate this problem I shall discuss briefly two examples from the United States. Susan Sontag, who traveled to Hanoi in 1968 and wrote and spoke with force and eloquence about her disgust and horror at America's war in Vietnam, quite dramatically revised her political stance in the early 1980s. Her views of her previous engagement have changed accordingly.[12] This fact is interesting from the perspective of Sontag's intellectual biography, but for the purposes of my study I need to know what her political actions and statements about the responsibility of intellectuals were during the Vietnam years.

Then there is the whole "Second Thoughts" movement of former radical intellectuals, who have moved far across the political spectrum and have come to question their earlier engagements. The best-known example is that of Peter Collier and David Horowitz, who served briefly as editors of the militantly antiwar periodical *Ramparts*. In 1989 they published a very interesting memoir, *Destructive Generation: Second Thoughts About the Sixties*. Neither of them speaks of a sudden moment of enlightenment when they abandoned their previous beliefs. They do admit, however, that "if there was one event that triggered our reevaluations (and those of others who began to have second thoughts about the Leftism of the Sixties), it was the fate of Vietnam."[13]

These political and philosophical shifts, which took place after the end of the Vietnam War, are quite fascinating and indicative of broader changes in American society that made the Reagan presidency possible and, at least from a short-

range perspective, so immensely successful. But if we focus
on second thoughts, we will lose sight of first thoughts and
attendant actions as they occurred in their contexts.

Thus the core of this book is based almost exclusively on
printed and primary sources, the antiwar writings of French
intellectuals between 1954 and 1962 and their American counterparts between 1964 and 1975. But even with this limitation,
the amount of available material is overwhelming. Herein lies
the fourth obstacle that I have had to confront. In a brilliant
synthetic article, which basically argues that intellectual history
is *en chantier* (still under construction), Jean-François Sirinelli
notes that anyone working on the history of groups of intellectuals runs the risk of the "syndrome of the miner." That
is, the volume of material that one has at one's disposal is so
immense that one is reminded of de Tocqueville's complaint
of 1853: "I was like the gold miner when the mine caved in
on his head: I was crushed under the weight of my notes and
I did not know how to get out of there with my treasure."[14]
It is easy, if somewhat discouraging, to illustrate why the "syndrome of the miner" is so contagious.

The period of the Algerian War has been seen, correctly
I believe, as "a golden age, perhaps the last golden age—of
reviews in France,"[15] and between 1954 and 1962 every major
French periodical devoted a large percentage of its pages to
issues connected with the war. In one of my principal sources,
the monthly *Esprit*—a journal that has generally taken since
its establishment in 1932 a liberal or left Catholic position—
211 articles, many of them long, on the Algerian War appeared between 1954 and 1962.

One way to combat the "syndrome of the miner" is, of
course, to rationalize production by the division of labor.
French scholars—more so than their American colleagues for
Vietnam—are passionately interested in the systematic and
detailed study of intellectual engagement during the Algerian
War. They have, wisely in my opinion, decided to work co-
operatively. An extremely able group of social scientists associated with the Institut d'histoire du temps présent [IHTP],

many of them too young to have directly experienced the Algerian War, is working on aspects of the general topic of "the Algerian War and French intellectuals."[16] The invaluable research already completed by these scholars is now appearing in print and will be discussed at several points in this book.

In the American case, while our Vietnam war was being fought, it produced an enormous volume of writing, including *Ramparts*, which flourished during the war, reaching a circulation of 300,000 in 1967 but surviving its end only by a few months, with the last issue appearing in August 1975. There was also a huge underground press. In a 1970 study of this "journalistic and social phenomenon," Robert Glessing listed 457 underground newspapers, with a circulation approaching 5 million in 1969.[17] In addition, the Vietnam War inspired professed scholarship, open polemics, novels, poetry, and every other imaginable type of writing. The bibliography did not stop growing when the war ended in 1975, and the volume of newly published materials shows no sign of abating. For example, the 1982 edition of John Newman's annotated bibliography of Vietnam War literature contains 226 entries; the second edition, which lists materials published through 1987, includes 752 entries.[18] No single scholar could even begin to acquire a command of this vast literature, and so I have had to be selective. One of my most valuable sources is the *New York Review of Books* (henceforth *NYRB*) which alone published 262 articles on the Vietnam War between 1964 and 1975, reflecting its obsession with that conflict.

What an *NYRB* author called in September 1986 "The War That Won't Go Away"[19] is intensively studied in our universities, with more than three hundred courses on aspects of the Vietnam era offered during the 1985–86 academic year.[20] Four years later, interest in the Vietnam War showed no sign of waning, with estimates of courses on it taught in American colleges and universities running as high as four hundred. For example, the most popular course at the University of California at Santa Barbara in the spring term of 1990 was "Religion and the Impact of the Vietnam War."[21] And in

August 1990 Jonathan Mirsky entitled his *NYRB* article reviewing eight recent books on Vietnam, "The War That Will Not End."[22]

With regard specifically to the role of intellectuals during the Vietnam War, little systematic work has been done. In 1974 Sandy Vogelgesang published an interesting but largely unnoticed book, *The Long Dark Night of the Soul: The American Intellectual Left and the Vietnam War*. Even though it addressed at some length the group of intellectuals associated with the *NYRB*, it was never reviewed in that journal. Vogelgesang's pioneering study also did not inspire further publication and research on questions of intellectual engagement during the Vietnam era, and since 1974 both journalistic and scholarly interest seems to have turned to other areas of inquiry. There has been sensitive and penetrating work on the Vietnam War as it is reflected in American literature,[23] but nothing similar to the collaborative study of French intellectuals and politics during the Algerian War undertaken by the Paris-based IHTP group.

In an extremely thoughtful essay in *Time* magazine, marking the tenth anniversary of the final North Vietnamese victory in 1975, Lance Morrow noted that "Vietnam, small and remote and poor, translated into an enormous presence in the American imagination."[24] This suggests that a crucial question to examine is the role of intellectuals as interpreters and perhaps even fabricators of the national imagination. Morrow also observed, tellingly I believe, that "Vietnam was a crisis of the American identity."[25] Intellectuals, at least since Alexis de Tocqueville, have busied themselves with the study of that identity. And finally, Morrow asserted that "the questions the war raised—in some ways still raises—were endless." Morrow then selected eight important questions that deal with a wide range of topics: diplomatic, military, political, and social. But only the seventh—"Were the soldiers of the peace movement representatives of a uniquely virtuous generation, the most idealist in history?"[26]—relates, albeit tangentially, to the engagement of the intellectual class.

For an article in the *Chronicle of Higher Education*, reviewing recent scholarship on Vietnam, Karen Winkler spoke with a number of the most respected scholars working in the field. She was told that current (1987) research on the Vietnam War "has yet to move beyond the debate that went on during the conflict itself." Based on her interviews Winkler drew up a list of four principal questions still in dispute. As it is worded, the third question, "Why did the United States get involved, and how can it justify its involvement?" might offer a future role for intellectuals if they would be willing to serve as government apologists. But the other questions are, like Morrow's, political, military, and diplomatic in nature. Because Winkler derived her set of questions from conversations with recognized experts, I believe that they reflect the direction of much of the future publication on the Vietnam War in this country. Hence I cite the other three here:

What was the real nature of the war: a nationalist struggle for independence, a civil war, or part of a larger, continuing conflict between communists and capitalists?

Who was responsible for beginning and escalating the war?

Why did the United States lose? Could it have won?[27]

I hope that this book, by filling certain gaps in our knowledge, will make answering Morrow's and Winkler's questions somewhat easier.

Also in 1987, George C. Herring, one of America's most eminent scholars of the Vietnam era, published in the *American Historical Review* an exceptionally interesting review article, "America and Vietnam: The Debate Continues." Herring commented that the debate over the war "has increasingly moved into the realm of scholarship."[28] Current work is more sophisticated, developing arguments in greater depth, though it remains polemical and we have a great distance to go. "Scholars are still sharply divided on Vietnam, and their divisions mirror deeper ideological and emotional conflicts among elite groups in the United States and across the world."[29] Our knowledge of the Vietnam War is incomplete and "profoundly confused." It may be many years, Herring believes, "before Vietnam moves out of the realm of politics and into the realm of history."[30] Another of my primary aims is to facilitate that move.

A fourth and final goal in writing this book is linked to its potential audiences. I anticipate that academics in France and the United States will find here arguments and conclusions to challenge and discuss. There also is another and larger group I would like to reach, which might be called the Vietnam intellectual generation or age cohort, not just the small number of intellectuals who actually fought in Vietnam. Few would quarrel with former Marine Lieutenant Philip Caputo's statement made in 1977, in the prologue to *A Rumor of War*—the powerful memoir of his service in Vietnam—that the Vietnam War was "the dominant event in the life of my generation."[31]

Whether that dominance will continue into the 1990s is uncertain and not empirically demonstrable. My own intuition is that it will, especially among the large surviving cohort of this nation's intellectual–professional class that came of age between roughly 1960 and 1975. I am convinced that Father Daniel Berrigan was correct in 1970, when he was in hiding from the FBI as a result of his own actions in opposition to the Vietnam War, in his belief that "this war has really forced a lot of people to stop and think about all sorts of things—to the point that they may never again be the same, these people."[32]

This same hypothetical group is described ten years later in perceptive, indeed haunting, article by Martha Ritter published in *Harvard Magazine*, "Echoes from the Age of Relevance." Ms. Ritter talks about her class of 1970 a decade after graduation, citing a remarkable comment from a classmate:

> I am struck by the fact that those of us who were in college during the late 60's remain different from those in college generations which came before or after. Coming of age during the Vietnam war... gave us a perspective on this society and our relation to it which, while probably no wiser than that of others, is unique. Even while many of us have followed traditional [career] paths,... we have maintained a certain distance, a feeling of being in some ways outsiders to this society in which we are now adults.[33]

If this statement is accurate, if it captures the sentiment of an intellectual generation, and I am persuaded that it does, I should like in my book to speak to that generation. In recovering and recording some "echoes" from the American "age of relevance," I hope to show that they themselves are in many ways echoes from another generation in another country.

1

Background to
Intellectual Engagement:
Historical Comparisons

The principal goal of this chapter will be to provide a historical
context for the focused study of intellectual engagement during the Algerian and Vietnam wars. Before launching into
this analysis we need to establish chronological parameters for
the events we are comparing, as there is disagreement even
about dates. One can argue that the French conquest of Algiers in 1830 and the subsequent colonization and economic
development of the region laid the groundwork for an inevitable struggle for Algerian independence. A reasonable
beginning date for the Algerian War, which some scholars
prefer, is May 8, 1945, with the spontaneous uprising in Sétif.
News of the German surrender in Europe, coupled with hopes
generated by the new United Nations Charter and the creation of the Arab League, led to rioting that spread from the
town of Sétif to the countryside. About one hundred Europeans were killed, and an undetermined number of Algerians,
conservatively estimated at seven thousand, were executed
during the savage military repression that followed.

For the purposes of this work I shall follow the narrower
chronology that dates the Algerian War from November 1,
1954, the outbreak of the carefully planned and precisely
timed insurrection organized by the Front de libération na-

14

tionale (FLN), which itself had been formally established only a week earlier. I have chosen as an ending date July 3, 1962, when Algeria officially became an independent nation. (The French themselves cannot agree on when to commemorate the termination of the Algerian War, some advocating March 18, 1962, when the Évian accords that granted Algerian independence were signed, others March 19, when the cease-fire was declared.)

Scholars sometimes call the American war in Vietnam the "second" Vietnam war, to distinguish it from France's own Indochina war which lasted from 1946 to 1954. The shadowy beginnings of our war in Vietnam may be traced at least to early 1956—when the United States assumed full responsibility for training the South Vietnamese army, taking over from the French—and probably earlier. In his excellent historical survey George C. Herring makes a good case for his choice of title, *America's Longest War: The United States and Vietnam, 1950–1975*.

Widespread public recognition, however, that the United States was involved in a conflict that was a war in everything but name, and significant intellectual response to that involvement, did not begin until 1964. Therefore, the term Vietnam War will here refer to the period of open hostilities beginning precisely with the adoption of the Gulf of Tonkin Resolution on August 7, 1964. This controversial resolution was passed by a vote of 88 to 2 in the Senate and unanimously in the House of Representatives. It was drafted in response to a presumed North Vietnamese attack on the American destroyer *Maddox*, which has never been satisfactorily documented and probably never took place.[1] But the resolution gave President Lyndon Johnson a blank check to move toward a full military commitment in South Vietnam.

As John Talbott notes in his valuable history of the Algerian War, there were remarkable similarities even in the way that both wars escalated after legislative votes that seemed to indicate massive national support. The Special Powers Law of March 16, 1956, was passed by a vote of 455 to 76 in the French Chamber of Deputies, an amazing majority given the usual close votes during the Fourth Republic. This legislation gave Premier Guy Mollet, his resident minister Robert Lacoste, and the French army free rein in Algeria, essentially removing any democratic controls over the manner in which the rebellion was to be suppressed. This "enabling act," writes Talbott, "was to the Algerian War what the Tonkin Gulf Resolution of 1964 was to the American War in Vietnam."[2]

A persuasive claim could be made that our Vietnam war was over on January 27, 1973, when the Paris cease-fire agreements were formally signed. The last American combat troops left Vietnam on March 29 of that year. Because the (Nguyen Van) Thieu government would never have survived for more than two years without massive American aid and because it provides a greater symmetry with the Algerian War, I prefer the more commonly accepted ending date of April 29, 1975, when U.S. Ambassador Graham Martin's helicopter lifted off the roof of the U.S. embassy in Saigon.

Similarities

Striking—one might want to use the word *painful*—similarities between the manner in which the Algerian and Vietnam wars were actually conducted once under way were pointed out long before the outcome of the Vietnam War was decided. To perceive patterns it was not necessary to wait for the passage of time to provide a cushion of objectivity or even for the Paris cease-fire agreements of January 1973.

As early as December 1964 a parallel from an antiwar perspective was drawn in the *New York Review of Books*. D. A. N. Jones reviewed a work by Lieutenant Pierre Leuillette entitled *Saint Michael and the Dragon: Memoirs of a Paratrooper*. The book had been published four years earlier in French and is an account of Leuillette's service in the Algerian War. The paratroop units with their green berets and leopard camouflage uniforms were the toughest and most brutal in the French army. (Identical uniforms were worn by the American Special Forces in Vietnam, and the physical resemblance between the two elite forces was noted at the time.[3]) Under General Jacques Massu the paratroopers took charge of the "pacification"[4] of the city of Algiers in 1956. Leuillette's memoir had been quite controversial and, like many contemporary books dealing with the Algerian War, had been seized by the French authorities after publication. Leuillette openly admitted that the French forces employed methods of interrogation and retaliation, writes Jones, "for which German war criminals were universally execrated and finally hanged."[5]

Jones goes on to observe that the Algerian War was almost as savage as the campaign that had been waged on behalf of the South Vietnamese dictator Ngo Dinh Diem. Leuillette—as did General Massu himself in a memoir published in 1971—calmly related what French antiwar intellectuals had been ardently claiming since 1955, that torture had been widely employed by France's army in Algeria, especially by the paratroops. Massu's blunt and reckless narrative—he tried out electrodes on himself to see whether he could tolerate the pain and was pleased that he could—is perhaps less interesting than the firestorm of criticism it aroused.[6]

By the end of his tour of duty in 1957 Lieutenant Leuillette did feel himself "becoming vile," and D. A. N. Jones believes that his mood, insofar as it was symptomatic, played a role in France's well-deserved defeat in Algeria. "The same disgusted mood makes the Vietnam war impossible to support."[7] Already in 1964, news photographs were being shown in England of "free world" tortures at work in Vietnam. Jones, a British novelist and critic, felt, and his prediction proved correct, that it would be difficult to gain European support for the American war effort in Vietnam. Leuillette's account, Jones thought, "preaches the lesson which it took him too long to learn, that there are certain ways of hurting which are not tolerable, which will strip you of your manhood and your will to win."[8]

Hence the issue of torture is raised, three months before the first marine combat units came ashore near Danang and seven months before President Johnson's July 1965 decision to authorize a massive increase in troop strength in Vietnam, "the closest thing to a formal decision for war in Vietnam."[9] In addition, the Nuremberg parallels are drawn for Algeria and by clear implication for Vietnam. Whether France itself might face the "bar at Nuremberg"[10] as a result of its actions in Algeria between 1954 and 1962 was an extremely painful subject of debate for French intellectuals and was laden with irony, as memories of sufferings endured during the Nazi occupation were so fresh. Later, the legitimacy of applying the precedents believed established by the Nuremberg trials of Nazi war criminals to the United States' involvement in Vietnam was to become a topic of passionate disagreement among American intellectuals.

These core debates and others that spun off from them are intimately related to, and in some cases directly inspired, the intellectual engagement that will be studied in later chapters. But first the broader historical context in which these debates took place will be examined in somewhat more detail, with specific attention paid to parallels between the two wars that can convincingly be drawn. No attempt will be made here to search for deeper reasons for those parallels, as such an inquiry would become a distinct research enterprise and another book. What is important to our concerns is that many observers, both during and after the Vietnam War, have believed that there were significant resemblances between the Algerian and Vietnam wars. As we shall see, that belief affected the forms of intellectual engagement and the attitudes and morale of *engagé* intellectuals during the Vietnam War. Important differences between the two wars will also be noted, as the similarities, superficially at least, are so striking that there is a temptation to find some kind of mysterious identity or cyclical repetition in the manner in which they ran their course. An American who lived in France during the Algerian War told me in 1989 that in his view Vietnam was a "xerox" of Algeria. If he was correct, the machine was not in good working order, and the copy that emerged was smudged in significant places.

A second reason to provide some of the historical background against which the intellectual engagement played itself out is that in both countries both wars have lost their historical immediacy, especially for younger generations. This is less surprising in the case of France and Algeria, given the extra decade that has elapsed, but it is a truth hard to grasp for many Americans who came of age during or before the 1960s, because the Vietnam experience is seared into their memories. When he was invited in 1981 to offer an undergraduate course on the Vietnam War, the political scientist and former State Department official Anthony Lake was somewhat taken aback and asked why such a course was needed. He was told, " 'The students don't remember it.' I would be teaching the seminal foreign policy event of my lifetime *as history*."[11] (Lake had been an aide on Henry Kissinger's staff and, after the invasion of Cambodia in April 1970, had resigned in protest.)

The Global Perspective

On a world-historical level, both bitter wars have been widely perceived as episodes in an ongoing, not yet fully completed, process of decolonization. For Jean-Paul Sartre and other radical Western intellectuals who, beginning in the late 1950s became enamored of the "Third World," decolonization was the most significant historical development of the second half of the twentieth century.[12] This view crossed ideological boundaries in a way that has been extremely rare in French intellectual history. It was shared, for example, by a moderate

like the eminent ethnographer Paul Mus, who in 1961, before
the war ended, was convinced that we were witnessing a "universal process of decolonization," and by the very conservative historian and author Robert Aron (not to be confused
with the more famous and somewhat more centrist Raymond Aron).[13]

Even though South Vietnam was never, strictly speaking,
an American colony, drawing such a broad parallel seems
quite reasonable and has been done by others.[14] But once
noted, this parallel is not especially enlightening, for it provides no particular reasons for singling out the struggles of
the Algerian and Vietnamese peoples from those of a large
number of other former colonized populations that in the
past forty years have acquired full political independence.
Rather, the similarities between what Stanley Hoffmann
viewed as "spectacular traps" for France and the United
States[15] become much more interesting when we turn our
attention specifically to political, diplomatic, and military comparisons.[16] There will inevitably be some overlap, but for clarity and coherence it should be possible to separate our
comparative analysis into these three categories.

Political Comparisons

In his brilliant work of instant history, *L'Algérie et la république*,
completed only two months after the May 13, 1958, military
rebellion in Algeria that brought General Charles de Gaulle
back to power, Raymond Aron argued that the Fourth Republic did not die from alcohol abuse or ministerial crises
or the parliamentary game; instead, "it died from colonial
wars."[17] With a very slight change in terms, the same statement could be applied to Lyndon Johnson's political defeat
and decision ten years later not to seek a second full term of
office.

This political parallel is, in my view, almost uncanny, and
I have not seen it discussed elsewhere, though because of the

enormous volume of material on the Vietnam War, this observation by some other writer may have escaped my attention. The comparison may be drawn out in some detail, in
that there were changes of regimes after four years of war in
both countries, with General de Gaulle's capitalizing on the
confusion generated by the May 13, 1958, coup to establish
the Fifth Republic with a constitution much more to his liking
than that of the Fourth. Richard Nixon was elected president
in November 1968. In each case a more conservative government than that under which the wars had begun, after long
and arduous negotiations often denounced as fraudulent by
impatient peace advocates, ultimately made peace. The agreements involved dramatic concessions, if not full capitulation,
to opponents that were much weaker militarily though not
politically.

By 1965, thoughtful observers were making specific political comparisons, pointing out similarities and differences
nearly eight years before the end of the formal American
military involvement in Vietnam. Christopher Lasch, who attended the important international teach-in held at the University of Michigan in September 1965, reported in *The Nation*
that he heard many references to what was assumed to be a
fact, that France had managed to withdraw from Algeria without losing its international prestige. None of the speakers
identified what Lasch believed to be obvious differences between the French case and the American: France was fortu-
nate to have a "national hero" in power, who "coupled
withdrawal with an appeal to glory." President Johnson did
not possess such charisma and was politically vulnerable as a
Democrat dealing with cold-war issues. Lasch thought, with
remarkable prescience, that Americans might have to wait for
a Republican president, "another Eisenhower," to end the war
in Vietnam and feared that by that time little of Vietnam
would remain.[18]

In September 1966 the historian and biographer Ronald
Steel visited Washington after having been absent from the

capital for two years. His first and deepest impression was
that Washington had become a city "obsessed by Vietnam."
There was virtually no other subject of conversation, whether
in a social gathering or a private discussion. Within official
Washington Vietnam had become, Steel thought, "not so
much a place as a way of thinking." The administration had
not been able to explain this war in terms that "could reconcile
it with traditional American values." As a result, Steel believed
that much of the nation's intellectual community had already
moved into opposition. Steel, who had begun his career as a
diplomat and spent time in Europe during the "long agony
of the Algerian War," did not feel out of place in the Washington of 1966. He found "the same impassioned commitment
by government officials, the same promise that the fighting
was in its 'last quarter hour,' the same baffled acquiescence
by the population, the same revolt of the intellectuals and the
same gradual erosion of confidence by the people in their
government."[19]

Many other political parallels were drawn while the Vietnam War was continuing, some of them quite enlightening
and convincing. Writing in 1968, John Gerassi linked Premier
Guy Mollet and France's future President François Mitterrand
to President Lyndon Johnson, an idea that would not have
been particularly attractive to any of the three politicians.
Gerassi cites Mollet's shout to the angry *pied-noir* crowd in
Algiers after he had been pelted with rotten fruit on February
6, 1956, the famous "Day of Tomatoes," when he changed
his mind about negotiation with the FLN: "France will fight
in Algeria and she will stay." Such rhetoric offered, Gerassi
thought, "a true inspiration to LBJ."[20] For many on the Left
who had supported him earlier, Mollet became the recipient
of a real aversion, even hatred, a "reborn Tartuffe, a paragon
of deception and betrayal." This antipathy crossed the Channel and was shared by many of Mollet's fellow socialists in
England. The British Labour party's weekly *Tribune* published
in April 1957 extracts from the antiwar brochure "Draftees

Bear Witness" (which had been banned by the Mollet government) under the title "He Disgraces the Name of Socialism."[21]

When one recalls the tremendous success of David Levine's political cartoons, such as the famous 1966 drawing of President Johnson with his abdomen bared and a map of Vietnam in place of the scar from his recent gall-bladder operation—which he had once proudly displayed to a surprised group of journalists—or that of Barbara Garson's savage satire of Johnson, *MacBird!*, also from 1966, and the favorable comments that the play received by such critics as Dwight Macdonald, Robert Lowell, and Robert Brustein, one may conclude that the parallel between Mollet and Johnson is not fanciful after all.[22] In 1980 with the benefit of hindsight John Talbott agreed with Gerassi: "Like Lyndon Johnson, Mollet became identified with a war policy that ultimately divided the country, tore apart his own party, and earned him the enmity of erstwhile political allies."[23]

Gerassi also claimed that a further inspiration for President Johnson was François Mitterrand's challenge to the FLN, made two years later, that is, in 1958: "Algeria is France. From Flanders to the Congo, only one law, only one nation, only one parliament. That is the constitution and that is our will. The only negotiation is war." This is somewhat unjust to Mitterrand, however, as the citation was in fact taken in a garbled form from a speech made on November 12, 1954, less than two weeks after the Algerian insurrection began, when Mitterrand was minister of the interior in the Mendès-France government.[24] Mitterrand's relative slowness in acquiring the conviction that Algeria would have to be granted its independence is widely recognized.[25] But in fairness to President Mitterrand it should be noted that early in 1955 he helped reduce the practice of police torture in Algeria by sending back to France certain inspectors who were known for their brutality. By 1957 he was perceived by the Right to be willing to abandon Algeria and was a target of death threats by fanatic paratroop officers. In 1961 his Paris residence was bombed with plastic explosives during the wave of terrorism mounted by the Organisation Armée Secrète (OAS), the secret paramilitary organization formed as part of the last-ditch effort to maintain Algérie française.[26]

An example of another type of parallel, though less convincing and even fanciful, may be found in Andrew Kopkind's April 1968 discussion of the mood of optimism and the belief that the war would soon end, which briefly followed the New Hampshire primary and President Johnson's announcement that he would not seek reelection. "The entrance of [Robert] Kennedy was the symbolic reappearance of the old hero who had been waiting in his Northern Virginia Colombey-les-deux-Églises [General de Gaulle's country home] for the magic moment."[27] A parallel that was amusing at the time it was drawn became tragic two months later with Robert Kennedy's assassination.

A third type of parallel, wishful and moving, if later proven inexact, is eloquently illustrated in Henry Steele Commager's plea for amnesty for deserters and draft resisters. Writing in April 1972, when the Vietnam War was winding down, Professor Commager reminded readers of the *NYRB* of the ancient definition of *amnesty*, from the Greek, meaning "amnesia" or "erasing from memory." After citing a number of historical examples of other nations that had been liberal with amnesties for activities judged disloyal by those in power, Commager writes: "It is perhaps even more pertinent to recall that that great soldier and statesman, General de Gaulle, proclaimed a general amnesty to almost all those who had resisted—even by arms—the government of France during the Algerian crisis."[28] The amnesty ultimately offered by President Gerald Ford, however, was quite limited.

Diplomatic Comparisons

From the perspective of diplomacy, an extremely early critique of the negotiation process was made by the French jour-

nalist Jean Lacouture. Lacouture was scornful of the great
"New Year's diplomatic offensive" unleashed at the end of
1965 by Secretary of State Dean Rusk and his associates, an
event that "treated American and international opinion with
a mixture of negligence and hypocrisy which reminds us of
the most bitter hours of our wars in Indochina and Algeria."
More specifically, Lacouture believed that American leaders
were at such a point of unrealism and blindness that their
behavior made one think of his nation's leadership just prior
to the Suez crisis of November 1956. Lacouture did point out
an interesting difference. Unlike General de Gaulle, who
wanted to negotiate only with "those who were fighting," and
not the Algerian provisional government in exile, Lyndon
Johnson was unwilling to deal with the South Vietnamese
combatants (usually termed the Vietcong in this country).
What may happen, according to Lacouture, is that Johnson
will discover that the Vietcong emissaries "have been pro-
moted to the rank of cabinet ministers."[29]

Many similarities and also important differences have been
found in the processes by which peace was finally made. In
1969, uncertain over how to terminate the war that Lyndon
Johnson had bequeathed to his administration, President
Nixon "was calmed about Vietnam with stories of the tortuous,
deceptive extrication from Algeria accomplished a decade earlier by Charles de Gaulle."[30] The raconteur, as one would
expect, was Henry Kissinger, his national security adviser and
later secretary of state.[31] Kissinger's own observations regarding this matter are quite self-revealing:

> Many urged us to "emulate de Gaulle"; but they overlooked
> that it took even de Gaulle four years to extricate his country
> from Algeria because he, too, thought it important for
> France to emerge from its travails with its domestic cohesion
> and international stature intact. He extricated France from
> Algeria as an act of policy, not as a collapse, in a manner
> reflecting a national decision and not a rout.[32]

Readers will note Kissinger's effort to indicate a close simi-
larity, if not an identity, between Nixon's approach to nego-

tiations (and thus that of Kissinger, the president's chief foreign policy adviser) and that taken by General de Gaulle.

Critics were quick to point out in 1973 the inexactness of this parallel, that the Paris peace accords did not produce full disengagement and that there was "no casting off of the South Vietnamese now that obligations have been fulfilled, no withdrawal approaching de Gaulle's withdrawal from Algeria."[33] However, some observers believed that on a deeper level the parallel was understood to be much closer, even before the settlement of 1973. According to this perspective, Nixon and Kissinger—in the case of the latter possibly as early as 1968— were aware that the GVN (the usual American abbreviation for the South Vietnamese regime, which will be used henceforth) would collapse and were, for internal political reasons, simply buying time by shoring up the Thieu government. In 1971 both Daniel Ellsberg and Stanley Hoffman foresaw the possibility of a reunited Vietnam under the rule of Hanoi and that there might be a "decent interval," or "face-saving time lag between our departure and their triumph."[34] The complete American withdrawal, of course, came only after an interlude of twenty-eight months, whether or not face-saving, depending on one's political views, with the rout of the South Vietnamese army in the spring of 1975.

Once the Vietnam War had finally ended and scholars could examine both wars with the benefit of hindsight, other interesting diplomatic parallels were noted and important differences discussed thoroughly by such writers as John Talbott and Alistair Horne. In his detailed and extremely useful history, *A Savage War of Peace: Algeria 1954–1962*, Horne analyzes the failed Melun peace negotiations of June 1960, when emissaries of the Algerian provisional government in exile (known as the GPRA) came to France. Horne points out that even de Gaulle was fooled, that he had in a de facto sense recognized the GPRA as a negotiating partner. The GPRA got a "first foot in the door which de Gaulle would never be able to dislodge." This diplomatic success offered "a lesson that was certainly not lost upon the negotiators of North Vietnam in their tediously protracted appearances in Paris over a decade later."[35]

Horne presents an interesting and persuasive interpretation of the Bizerta crisis of July 1961, which if remembered at all is not normally viewed as having had a significant impact on the course of the Algerian War. Tunisian President Habib Bourguiba, for his own political reasons, decided to attempt to force an early evacuation from the key naval base at Bizerta that had been leased to France after Tunisia was granted independence in 1956. De Gaulle ordered a harsh counterattack that completely defeated the Tunisian army at a cost of twenty-four French lives.

> By riposting with such crushing force, de Gaulle may well also have had at the back of his mind one final display of military might to impress the F. L. N. all the way to the peace table [and to get terms short of full independence]— much as President Nixon was to launch his all-out bombing of Hanoi on the eve of peace over a decade later. But, if this was his aim it was equally counter-productive.[36]

After Bizerta, de Gaulle lost the support of the one moderate regime that would have exerted influence on the GPRA to accept compromise peace terms, and France was once again sharply condemned at the United Nations. Likewise, the Christmas 1972 bombing of Hanoi produced appalling destruction and tremendously increased casualties on both sides but did not pressure the North Vietnamese into accepting terms more advantageous to the United States and the GVN.

Pursuing the comparison further, Horne observes that by January 1962 de Gaulle had recognized that there was no negotiating option but total independence, and he was determined to rid himself of the "Algerian problem" as swiftly as possible. "Like Henry Kissinger with Vietnam in 1973, he was a man in a hurry."[37]

Finally, in his conclusions regarding the role of de Gaulle in the Algerian affair, Alistair Horne argues that the general

suffered from the lesson not learned by Kissinger in Vietnam...that peoples who have been waiting for their independence for a century, fighting for it for a generation,
can afford to sit out a presidential term or a year or two in
the life of an old man in a hurry.[38]

Thus we have circled back to the global perspective with which
we began our comparative analysis, while viewing decolonization within the specific historical contexts of Algeria and
Vietnam.

Military Comparisons

Certain compelling similarities in the manner in which the
Algerian and Vietnam wars were fought have already been
discussed in the introduction to this chapter. We shall now
examine military matters in further detail, paying some attention to the effects of the wars on the indigenous populations.

The interweavings between the two wars are immensely
complicated, and although many writers have noted one or
another parallel, they have never been systematically analyzed. The similarities run from the broad strategic consid-
erations of the political leaders and the commanding officers
to the basic tactics and methods of combat, staffing, and main-
taining order and morale in the armies. For example, if a
French draftee's father had been killed in World War II, he
was exempted from service in Algeria. The Selective Service
System used the same principle for the American army in
Vietnam. In both wars, the threat of removal of draft deferments from student activists was used as a way of quieting
dissent.[39] And there were essentially identical racist attitudes
on the part of the common soldiers.

Though the American arsenal was, of course, vastly more
diverse and technologically sophisticated, there were similarities in the types of weaponry and the manner of their utili-
zation. Napalm and helicopters were used in both wars, and

disposing of prisoners by throwing them out of helicopters was commonly practiced in Algeria and Vietnam. The maximum troop strength of both expeditionary forces numbered approximately one-half million, before ceilings were set by the political authorities, although the generals asked for more. Roughly three million soldiers from France and three million from the United States did tours of duty in Algeria and Vietnam. Given the fact that France's population was and is about one fourth that of the United States, the burden on the French age cohort susceptible to induction into the army was much greater. Draftees served in both wars, as they had not during France's war in Indochina, which was fought by a combination of volunteers and mercenaries.

On the level of individual leaders, we may cite the example of Air Marshal, Premier, and later Vice-President Nguyen Cao Ky, undoubtedly the most flamboyant and controversial of the South Vietnamese governing elite. Ky is perhaps best remembered for his October 1964 statement that his greatest hero was Hitler and that "we need four or five Hitlers in Vietnam." Ky attended aviation school in France, completing his training in September 1954, and flew with the French air force. Last, there is good, though not definitive, evidence that Ky served in Algeria as a fighter pilot carrying out missions against the FLN in the early months of the uprising.[40]

The commanding officers and the politicians who were their nominal if not real superiors in both wars suffered from an identical blind and persistent optimism. In the French case the stock phrase was *le dernier quart d'heure* (the last fifteen minutes); in the American, "the light at the end of the tunnel." One example from the Algerian War will suffice. In an interview given on April 22, 1958, less than a month before the military coup that led to his removal and the establishment of the Fifth Republic, Robert Lacoste stated, "We are just at the moment of reaching our goal" (*Nous touchons au but*). The Tunisian frontier is sealed, Lacoste stated, and if the FLN tries to bring arms in through the Moroccan border, we can stop them there, too. All the cities are under control, except for Constantine, and he planned to send a regiment of paratroopers there. Then we shall see "a French victory." Lacoste did admit that there were some regions in the countryside that still had to be pacified, but "we shall take care of them."[41]

The casualty estimates, termed *head counts* or *body counts*, that were prepared by the information officers of both armies were notoriously unreliable,[42] and combat officers such as Lieutenant Philip Caputo were well aware of this. Caputo was temporarily assigned as the officer responsible for verifying the body count and the resultant kill ratio for his unit. One common practice was to count any dead civilians as Vietcong, which naturally made the kill ratio higher and victory appear nearer at hand.[43]

In language that was repeated again and again almost verbatim by his *confrères* who covered Vietnam—David Halberstam, Michael Herr, Gloria Emerson, Bernard Fall, Frances FitzGerald, and so many others—the astute French reporter Albert-Paul Lentin wrote (after spending four months in Algeria at the end of 1958 and early 1959, when Lacoste's successors were making the same kinds of statements): "One will not easily unburden the heads of French generals and ministers of the chronic conviction, perpetually denied by facts, that 'victory is a rifle shot away.' Their illusions are tenacious, their optimism, while completely irrational, is also impregnable, and resists every tempest."[44] Again like his American colleagues a decade later, Lentin found that the so-called intelligence services in Algiers were flooding Paris with reports claiming that the FLN's morale was down, that its recruitment had ceased, and that it was ready for catastrophe. The agents were obviously taking "their desires for realities." In Lentin's view the war was stalemated militarily, and he foresaw—correctly—a long struggle ahead.

To pursue further this parallel with a direct comparison, Alistair Horne pointed out that the French officer corps could see with "agonizing clarity" that they were at certain points winning the war militarily. In this, I would add, they were probably more accurate than their American counterparts, given the fact that by the end of 1958 they were indeed able to seal the frontier with Tunisia, essentially halting the flow of arms to the FLN. "But (not unlike the American commanders in Vietnam) it was not given them to perceive that, at the same time, their chances of winning the war politically and on the wider world stage were growing every slimmer."[45]

Delusions were not the sole property of the high commands. Officers with the rank of lieutenant through colonel in both armies became "counterinsurgency cultists," developing a "centurion complex."[46] Not only did the paratroop and Special Forces units wear identical uniforms, but they also formulated similar theories of counterrevolutionary warfare, imagining at least that they were borrowing from Mao Zedong and Vo Nguyen Giap and Ho Chi Minh, whose works they read while in the field. The very notion of counterinsurgency as elaborated in Vietnam, with the help of teams of researchers primarily from Michigan State University, was borrowed from the elite of the French officer corps who used it in Algeria. In both wars, through "pacification" programs, this new breed of more educated soldiers came to believe that they had the power to enact social reform.

It is not surprising that toward the end of the Algerian War the OAS newsletter designed to propagandize the officer corps of the French army and convince it of the justice of the cause of Algérie française was entitled Les Centurions.[47]

As the French war correspondent and historian Bernard B. Fall, one of the most brilliant and prolific observers of the early years of the Vietnam War, pointed out, the centurions in both wars were convinced that "we're here to defend the free world even if it is too fat and sloppy and rich to help us; even if it does not want to be defended."[48]

Journalists covering both wars recognized that the efforts of the Special Forces to gain the loyalty of the peasantry were delusory and futile. What Albert-Paul Lentin wrote after a tour in November 1958 of the Bled, the Algerian countryside, applies exactly to rural South Vietnam a decade later. He interviewed a number of junior officers who worked as teachers and nurses helping the villagers. "Only during the day, to be sure, for at night Algeria belongs to the FLN."[49] (Simply reverse the initials to NLF, the National Liberation Front, commonly called the Vietcong by the Americans.)

In actual combat areas, during "search and destroy" missions, it was always difficult to distinguish friend from enemy, especially because the villagers in both wars were usually sympathetic to the rebels in their midst.[50] This, of course, led to increased reprisals and civilian casualties and ultimately massacres such as that at Mylai.[51]

To accomplish the ever-elusive goal of "pacification,"[52] similar techniques were employed in Algeria and Vietnam, including the relocation of vast civilian populations. Estimates vary wildly; using conservative figures, nearly two million Algerians and one and a half million South Vietnamese were uprooted (in official language, "evacuated") into what were euphemistically called *camps d'hébergement* (shelter camps) or *regroupement* (transit) camps. Some writers, however, have not hesitated to identify these sites as concentration camps and to use the highly charged word *deportation* to describe the resettlement programs.[53] Then, once the areas were cleared, anyone left in a *zone interdite* (forbidden zone)—"free-fire zones" in Vietnam—was deemed to be a *fellagha* (an FLN irregular) or a Vietcong, as the case might be.[54] Napalm was used with little or no discrimination in these regions, which led to the massacre of untold numbers of innocent civilians who simply happened to be in the wrong place at the wrong time. It is not difficult to understand Robert Crichton's assertion in *Authors Take Sides on Vietnam* (1967) that any society capable of using such weapons as napalm, and becoming committed to them, "must eventually succumb to the sickness of its own violence. France found this out in Algeria. Something has gone out of France."[55]

In addition, the terrible impact of drugs on the American army in Vietnam, the rapid increase of opium smoking between 1967 and 1970, the ready availability of cheap and pure heroin, and the resultant epidemic of drug addiction in American society are unquestioned and widely publicized facts and have been discussed in an extensive literature.[56] What is less well known is that hashish was smoked in Algeria, that there were addicts in the paratroop units, and that some medics made a fortune selling morphine and maxiton (an amphetamine).[57]

In his powerful 1972 work denouncing General Jacques Massu for his now admitted authorization of the use of torture in the Algerian War, Colonel Jules Roy draws a parallel with the United States in Vietnam, in which our army also tortured "scientifically." In some cases even the same implement was used, the field telephone with electrodes attached to the genitals and other sensitive body parts of the victims.[58] This is the same "gangrene," and it now rots the American army. "Do not look elsewhere for the reasons for the defeat of America," writes Roy, three years before the fall of Saigon.[59] For the purposes of this comparative analysis, it is not necessary to elaborate further on the widespread and well-documented use of torture on all sides in both wars. But we shall return to it from the perspective of the intellectuals, when discussing their campaigns against torture.

Differences

Although there are many intriguing political and diplomatic similarities that emerge when one thinks comparatively about the two wars, the most powerful similarities are clearly military. But even in this area there were significant differences, for example, in the ways that auxiliaries and allies were used (the *harkis* in Algeria versus the technically separate and autonomous regular South Vietnamese army). Other military differences, especially those relating to the greater population

and economic resources of the United States, have already
been cited.

A good case could be made, however, that the dramatic
difference in geographic distance from the homelands and
the ensuing logistical difficulties were essentially erased by the
power of American technology. Just as a promising young
chef could be flown home on special leave for a few days from
his unit in Algeria to his native village to prepare a meal for
the visiting President de Gaulle in 1959, so could in 1970 a
young employee of CORDS (Civil Operations and Rural Development Support) make a quick round trip from his post
in Saigon to attend his sister's wedding in the United States.[60]

Among other important differences between the two wars
are those connected with ideology and religion. There was
obviously no "Arabo-Islamic" ingredient in the ideology of
the North Vietnamese and the Vietcong, and no official Marxism in the doctrines of the FLN, which was sharply, and jus-
tifiably, critical of both the Algerian and French communist
parties for their lack of support.[61]

Quite probably the most significant difference between the
two wars was the presence of a settler population of close to
a million in Algeria, which could claim roots in North Africa
going as far back as 1830, a much longer historical memory
and historical involvement in the region than that of the
Americans in Vietnam. Our presence there can be traced to
1945 at the earliest, but 1950 would be a more reasonable
beginning date. The *pieds-noirs'* passionate attachment to a
land that with some justification they viewed as their own
influenced in countless ways the course and the immediate
aftermath of the Algerian War, as the vast majority of this
dynamic and ebullient population returned to France, often
under appalling conditions, during the summer of 1962.[62]

Perhaps equally significant was the presence in France of
about 300,000 Algerian workers, who with their monthly contributions to the FLN and their government in exile, the
GPRA, whether voluntary or coerced, paid a high percentage

of the cost of the war. As Ali Haroun argues convincingly in his superb history of the Algerian War as it was fought in France, "it was the Algerians themselves who financed their war of liberation."[63] Haroun points out with equal pertinence that the GPRA's financial independence was closely related to the genuine neutrality that Algeria has been able to observe since independence, casting its lot with neither the Eastern nor the Western bloc, and also its freedom from influence from other Arab nations, to whom it had never been "constrained to extend a begging hand."[64] The Vietnamese rebels, both northern and southern, did not have the support of an émigré population and so had to rely on Soviet and Chinese aid, with all the strings that were and to some degree still are attached.

I would argue that these differences are of greater significance—at least in the ways that the wars played themselves out—than the juridical differences. Algeria was, according to French law, a part of metropolitan France, divided into *départements*, with elected representatives sitting in Paris in the Chamber of Deputies. Likewise, when the Americans began to arrive in Vietnam en masse, they tended to treat their South Vietnamese allies and the country itself as almost their property, certainly as a colony. Philip Caputo observed that his fellow marines who served with him in Vietnam "believed that the ground they stood on was now forever a part of the United States simply because they stood on it."[65] In fact, the Saigon government was barely consulted on major decisions such as the bombing of North Vietnam and the bringing of U.S. Marines to guard the air bases. After 1964 the American domination of South Vietnam was almost total. The United States "controlled the fighting, financed the country, and selected or influenced the selection of leaders down to the village level."[66] Many thoughtful South Vietnamese, including those with no sympathy for the Vietcong and the Communist regime in the North, viewed their governement as a puppet, and the U.S. ambassador was frequently referred to as "the governor-general"[67] (a position that had long existed in both French Indochina and Algeria).

Some of the differences in the ways that the wars affected the home fronts relate closely to the response and engagement of the intellectuals and will be discussed in later chapters. Suffice it to say here that the psychological impact of the Vietnam War on the American public was inevitably different. As John Talbott observes, television did not become a fixture in French households until the later 1960s, and thus Algeria was not the " 'living room war' that Vietnam later became."[68]

In this unavoidably selective presentation we have attempted to provide relevant background as a prelude to the comparative study of two extraordinary episodes in the history of intellectual engagement. We should be able to understand better why the eminent Catholic antiwar intellectual Jean-Marie Domenach, himself no extremist, could publish in October 1957 a moving article entitled "Culpabilité collective," (Collective guilt) in *Esprit*, the review he edited. Domenach begins by referring to certain scandalous events such as the "disappearance" in Algiers of the mathematician Maurice Audin (it was later discovered Audin died while being tortured by the paratroopers; his case will be considered in Chapter 3) and the journalist Henri Alleg's powerful account of the tortures he endured after being arrested, also by the paratroopers, in June 1957. Domenach notes that despite masses of incontrovertible evidence, no sanctions have yet been taken; "the hangmen are still on the job." He does not accuse any particular group or preach morality to others. The moral question, furthermore, does not apply to only a few executioners but to all Frenchmen.

> Beyond a few individuals responsible the hierarchy and who should be punished, it is the entire collectivity that bears the blame. By this I mean the entire nation, its institutions, its newspapers, its churches, all of us Frenchmen who allow 400,000 young men to sink into a war in which the search

for information, to be obtained no matter what the price, and the repression of an elusive terrorist movement have led to such excesses. We French all are responsible, and that includes those on the Left whose political impotence has allowed free rein for the actions of imbeciles and the blood-thirsty [*homes de sang*].[69]

We should also be able to understand why Susan Sontag could write a decade later, in 1967,

America has become a criminal, sinister country—swollen with priggishness, numbed by affluence, bemused by the monstrous conceit that she has the mandate to dispose of the destiny of the world, *of life itself*, in terms of her own interests and jargon.

America's war on Vietnam makes me, for the first time in my life, ashamed of being an America.[70]

We are now ready to turn to the intellectuals and to two cycles of their engagement.

2

Cycles of Engagement

Definitions

As someone who is professionally identified as an intellectual historian, I am frequently asked, "What is your definition of an intellectual?" and "What do you understand the 'engagement' of intellectuals to mean?" These are fair, if difficult, questions and deserve replies that go beyond Bernard-Henri Lévy's imaginary dictionary definition cited in the Introduction, and the briefly noted relationship between intellectuals and engagement. My responses should provide the necessary framework for both the theoretical analysis to be presented in this chapter and the case histories discussed in Chapters 3 and 4.

In the attempt to arrive at a satisfactory definition of the vaguely delimited social grouping known as "the intellectuals"—located in Western societies, the United States included, somewhere in the middle of the middle class—an enormous amount of ink has been spilled by self-proclaimed intellectuals and by anti-intellectual writers who refuse the appellation but probably still belong in the category.[1]

There is no need to review this debate here, as for our purposes a loose operational definition of the type proposed by eminent sociologists such as Talcott Parsons and Edward Shils will suffice. Intellectuals are defined by their more abstract and distantiated social role which sharply contrasts with almost all others in a modern society. Their function involves a certain kind of creativity, usually through the written word and dealing with ideas in some fashion, often applying ideas in an ethical way that may question the legitimacy of the established authorities. Sandy Vogelgesang makes this point very succinctly: Intellectuals are "the men and women of ideas who explore and challenge the underlying values of society. Theirs is a normative function: to prescribe what ought to be."[2]

There is an important delimiting factor that finds its way into almost all definitions and is widely accepted by both those who view themselves as intellectuals and those who would angrily refuse to be so labeled. Even if not strictly impractical and utopian, the manifold varieties of intellectual production do not have a directly measurable effect on society or great economic value in societal terms, and intellectuals (at least those whom Antonio Gramsci would term *traditional* intellectuals)[3] do not hold political power.

The corollary to this view, which is also commonly accepted, is that when an intellectual is "absorbed" or "coopted" into a corporation or a government post, he or she loses pure status and enters a hybrid social category, sometimes called *establishment intellectuals*.[4] Indeed, sociologists have often focused more on what intellectuals are not than on what they are. Edward Shils notes that "in private economic organizations, the employment of intellectuals in administrative capacities has been uncommon to the point of rarity. Nor have intellectuals ever shown any inclination to become business enterprisers."[5] As Shils puts it, there will always be a "tension between the intellectuals and the powers."[6]

I do propose to keep my working definition flexible and cast my net more widely than some. I would categorize a significant percentage of the professoriate and some journalists as intellectuals, as well as a substantial portion of the artistic community—painters, sculptors, musicians, and the like—who theorize in print about their creativity. (The claim of a commercial or pop artist to intellectual status, if one ever cared to make it, would be dubious at best.) University students thus would be understood as potential or apprentice intellectuals, though only a limited number would ultimately become full-fledged members of the guild.

Accepting the sociological definition just outlined leads to the obvious conclusion that a social subgrouping that can legitimately be categorized as intellectual has existed in some form at least since ancient Greece. In Christian Europe the intellectuals were the clerics, as Julien Benda argued in his passionate denunciation of modern intellectuals for their betrayal of their true idealistic vocation, *La Trahison des clercs* (1927).

It is, however, important to remember the historical context in which the noun *intellectuel*, demarcating a segment of society, became a part of everyday French vocabulary. This did not occur until January 1898, at the height of the Dreyfus affair, with the publication of Émile Zola's *J'accuse* and the manifestos and countermanifestos that followed.[7] The term *intellectual* crossed the Channel and then the Atlantic almost immediately and entered common English usage.

Hence when a group with a conscious self-identity calling itself for the first time *les intellectuels* emerged in France during the Dreyfus affair, it did so in order to take a political stance. There was, and perhaps remains, a symbiotic relationship between the intellectual and engagement. This indisputable fact produces a paradox: Just as the concept of the intellectual was being elaborated and the word coined, the temptation to abandon the activities generally understood as falling within the intellectual sphere was present and was irresistible, at least temporarily.

The French word *engagement* has had a variety of meanings, with as many as thirteen definitions listed in some dictionaries.

In the 1930s a new political definition arose quite sponta-
neously, probably in response to the catastrophic upheavals
of that decade. *Engagement* in this usage ordinarily referred
and continues to refer to political involvement by members
of the intellectual class—however broadly or narrowly defined
a social group that is widely viewed as not normally prone to
descend from the ivory tower into the arena. This view remains constant as historical contexts have shifted. In June
1990 the French intellectuals were denounced by an edito-
rialist in *Le Monde* for their scorn of political life, their "pur-
ism," which has indirectly helped the growth of the extreme
right National Front party of Jean-Marie Le Pen. In this dangerous situation once again the intellectuals are called to
engagement.[8]

An important corollary to this general definition is that
the intellectual or intellectuals in question are not pushed
down the stairs of the ivory tower and out into the streets.
True engagement cannot be coerced but is derived from reflection on the external political and social situation, and a
conscious and reasonably free decision to become involved.
Thus during the Second World War in Occupied Europe,
intellectuals were not ordinarily forced to join resistance
movements; that is, it was possible to remain prudently in the
background, and many did.

In English the noun *engagement* and in French the adjective
engagé used in this activist sense appeared first in scholarly
studies after their importation from France, thanks in large
measure to their popularization by Jean-Paul Sartre in *What
Is Literature?* (1948). By the 1970s, *engagement* had gained wide
enough acceptance to find its way into mass-circulation periodicals such as *Time* magazine, which in 1971 characterized
André Malraux as the "archetypal *homme engagé*, the intellectual man of action," and *The New Yorker* used in July 1990 the
same words to portray Malraux.[9] The syndicated columnist
George Will wrote in 1987 that "as an intellectual in politics,
[Victor] Hugo exemplified the modern ideal of 'engage-

ment.' "[10] Many would argue that this ideal has become tarnished, and in 1986 Richard Bernstein mused in the *New York Times* that for intellectuals it may now be *"passé* to be *engagé."*[11]

These selected examples suggest, if they do not demonstrate conclusively, that Americans, at least the literate classes, now have a broad if somewhat imprecise understanding of the concept of engagement. There is also a general awareness of the distinction between intellectual engagement and the workaday activities of writers, professors, artists, and others in related intellectual professions. Erica Abeel made this point clearly in the *New York Times* in 1979, when she worried that deteriorating conditions at the City College of New York had obliged concerned faculty to "substitute *engagement* for research."[12]

Parallels and Lessons—Real and Imagined

Were there any similarities in the intellectual engagement generated by the Algerian and Vietnam wars? Did American intellectuals learn from the French experience and if so what did they learn? These questions have rarely if ever been asked and have never been examined systematically. Hervé Hamon and Patrick Rotman's definitive study of the militant intellectual resistance to the Algerian War, *Les Porteurs de valises*, was published in 1979. The four years that had elapsed since the fall of Saigon would have allowed the authors of this extremely rich and detailed history, which conveniently reprints important documents, time for retrospective analysis and comparison had they wished to draw parallels. But Hamon and Rotman do not once mention America and Vietnam.[13] Indeed, the only work in English dealing with the French intellectual response to the Algerian War is Paul Sorum's *Intellectuals and Decolonization in France*, published in 1977, also after the end of the Vietnam War. Sorum does indicate in his preface his "personal dislike of ethnocentrism, colonization and warfare—sharpened by America's ordeal in Vietnam."[14]

This suggests that Sorum's personal views on Vietnam influenced his choice of subject and perhaps affected his generally sympathetic treatment of the French antiwar and anticolonialist intellectuals, but he makes no reference in the entire text to the American intelligentsia and Vietnam.[15]

Even more striking is the fact that in the superb collaborative volume *La Guerre d'Algérie et les intellectuels français*, published in November 1988, only one of the authors, Rémy Reiffel, in a very brief though intriguing aside, speaks of Vietnam.[16] In 1987 and 1988 Reiffel interviewed fifteen of the most eminent surviving French antiwar intellectuals, and one of them, Pierre Vidal-Naquet, the cofounder of the Comité Maurice Audin, told Reiffel that opposition to "the Vietnam War mobilized more people than did the Algerian War in France."[17] Vidal-Naquet's assertion, of course, comes through the filtering—or "decantation" as Reiffel would say—of memory and without having studied the Vietnam episode in depth. Professionally, Vidal-Naquet is an eminent classicist, several of whose works have been translated into English. Testing his claim in a precise quantitative fashion is certainly impossible. My intuitive conclusion is that Vidal-Naquet is correct if one does not take into account the fact that France had one quarter the population of the United States. To make the comparison legitimate one might multiply by four the number of engaged French intellectuals and come up with a number roughly equal to the number of American activists.

In the one full-length study of the American intelligentsia and the Vietnam War, completed after the cease-fire of January 1973 and published in 1974 before the final North Vietnamese victory, Sandy Vogelgesang already perceived significant parallels in two areas. She mentions, though only in passing, that by 1965 American antiwar intellectuals "stressed what they considered the farsighted statesmanship of French withdrawal from Algeria." Second, "they drew parallels between intellectual opposition within France and the United States, caused by Algeria and Vietnam respectively."[18]

These tantalizing statements are offered without documentation, and Vogelgesang does not indicate which intellectuals
or which parallels she had in mind.

In Chapter 1 we cited a number of examples that support
Vogelgesang's first assertion, and her second is also largely
corroborated by my research. Elsewhere in her text she does
mention one very telling illustration. She refers twice to the
famous "Call to Resist Illegitimate Authority," published in
October 1967 in the *NYRB* and the *New Republic* and widely
circulated thereafter. This petition, one of the most important
documents in the intellectuals' campaign against President
Johnson and the Vietnam War, had by June 1968 four thousand signatories, two thirds of whom were from university
faculties. It led directly to the establishment of the militant
antiwar organization called Resist.[19] The "Call to Resist Illegitimate Authority" was drafted in the spring of 1967 by Marcus Raskin and Arthur Waskow, who, Vogelgesang claims,
borrowed consciously from the "Déclaration sur le droit à
l'insoumission dans la guerre d'Algérie," commonly known as
the "Manifesto of the 121," of September 1960, signed by
Simone de Beauvoir, Jean-Paul Sartre, and many members
of France's intellectual elite of the time.[20]

This direct borrowing is documented through an interview
with Marcus Raskin, who in 1965 coauthored with Bernard
B. Fall the influential *Vietnam Reader*. Fall, who had covered
France's Vietnam war as a journalist and written two important books on it, was teaching at Howard University while
making periodic trips to Vietnam to observe and write about
our war. He was killed in 1967 in an ambush by stepping on
a land mine while on patrol with an American unit. Raskin
credited Fall with the original idea for the "Call to Resist
Illegitimate Authority." By 1966 Fall had lost any journalistic
neutrality he might have possessed and had become strongly
opposed to the war. As Raskin writes,

> His greatest anger was directed against the American left,
> their failure to become aroused over the torture of Viet-
> namese prisoners of war and the use of napalm. There

seemed to be no one to speak out against it as the French
did on Algeria. This turned me on to thinking about something similar to the "Statement of the 121."[21]

In his anguish over his inability to put a stop to what he
had seen during his visits to Vietnam, Bernard Fall may have
been a little unfair to his fellow intellectuals in his adopted
country, for as we shall see in Chapter 4, serious intellectual
protest had begun in the United States as early as 1965.

As the war and the protests escalated, American antiwar
intellectuals, while searching for techniques and models to
follow, not only believed that parallels existed between the
Algerian and Vietnam wars but also found inspiration and
hope from examples of antiwar engagement drawn from
France and Algeria between 1954 and 1962.

Father Daniel Berrigan, "the Holy Outlaw" and one of the
most notorious and effective of the antiwar intellectuals, had
studied in France in the 1950s, was well aware of French
Catholic engagement, and consciously used it as a model for
his own, even imitating the style of dress of the French worker-
priests.[22] In a journal he kept between 1965 and 1967, some
years before he served a sentence in federal prison in Danbury, Connecticut, Berrigan commented on the criminality of
our involvement in Vietnam: "Our situation is a great deal
like the Algerian war; priests then were silenced, and even
(since the French are less churchy than ourselves), thrown
into jail for protesting violence and terror and torture."[23]

Inspiration could be found in the example of French opposition to the Algerian War, even when the Americans drawing the parallels were completely wrong from a historical point
of view. In April 1967 Paul Goodman, the eminent social critic
and writer on education who had become a forceful and eloquent opponent of the Vietnam War, stated that the American students who burned their Selective Service cards, a
"crime" punishable by a maximum of five years in prison,
employed "as a model the similar extreme action of French
youth which did begin the withdrawal from Algeria."[24] This
is an exact quotation, and I simply have no idea of how Good-

man arrived at that notion. No contemporary observer or
participant on French soil, whether for or against the Algerian
War, no French historian or sociologist or political leader, or,
for that matter, student leader has ever made such a claim.
The only possibility that I can imagine is that Goodman had
some memories of the spontaneous disturbances created by
French draftees in the fall of 1955, when groups of soldiers
(not students), angry at having their terms of duty extended
and being sent to Algeria, stopped the trains taking them to
Marseilles for embarkation by boat to the war zone. Some of
these uprisings became serious riots and caused problems for
the authorities before they could be suppressed and the trains
rerouted. There was little support for these actions among
the populace, however, and none of the political parties, including the Communists, came to the aid of the rebellious
soldiers. The riots quickly subsided and were not repeat during the remaining seven years of the Algerian War.[25]

In an article published in May 1967 Goodman again drew
the Algeria–Vietnam parallel in more general terms. He
quoted a student draft resister who claimed that the goal of
the "We Won't Go" organization was to start a mass movement
of resistance to the Vietnam War, "the way the French got
out of Algeria."[26] There was some student protest in France
during the Algerian War, especially in its last three years, but
it was limited in scope and never developed into anything
approximating a mass movement.

Finally, in the wake of the October 1967 march on the
Pentagon in which he participated, Goodman, who died in
1972 and thus never saw the end of the Vietnam War, was
momentarily enthusiastic, believing that "there is a ground-
swell of American populism. The climate is beginning to feel
like the eve of the French withdrawal from Algeria, including
the same coalition of the young, the intellectuals, and the
Algerians (Negroes)."[27]

Although the real existence of the parallels that Goodman
and the American student leaders found is impossible to sus-

tain—in large measure because of flawed historical knowledge of what actually happened in France during the Algerian War—the fact that they were consciously drawn and give every indication of being firmly believed in is not. There is no reason to doubt their effectiveness as morale builders, for however inaccurately recalled in its details, the Algerian conflict was remembered as a victory of anticolonialism and the ultimate triumph of the forces opposed to that war.

If we focus on the antiwar intellectuals themselves and the types of their engagement and avoid premature evaluation of the effectiveness of their action, other much more convincing parallels will emerge. From this perspective and with the benefit of hindsight, we can discern a near identity between two fascinating episodes in the history of the intellectual class in the twentieth century. This identity overshadows and plays a role in reducing certain long-standing differences between the two intellectual classes, such as the fact that traditionally France's intellectuals had enjoyed greater prestige and had a more positive self-image.

Cycles of Engagement

The primary focus of this book is on the undeniable majorities of the intelligentsia in both countries who were opposed to war, majorities that grew as the wars progressed. Though exact measurement is not possible, especially because of the persistent uncertainty about whom to include under the rubric of intellectual, some percentage estimates will be given in later chapters. Almost all the scholarly and journalistic studies that have been published thus far have dealt with oppositional groups. An intellectual obviously does not have to be liberal, radical, or antiwar in order to be politically involved. Chapters 3 and 4 describe the counterengagements of the prowar or at least progovernment intellectuals, which indeed merit detailed comparative treatment in another full-length book.

After all, as Jean-François Sirinelli showed in the first serious examination of conservative engagement during the Algerian War, a true encephalogram of an intellectual class should also examine the right lobe of the brain.[28]

My research led me to conceive of these cases of engagement first in terms of varieties, with no clear connections between each type of activity, then as separate but distinct patterns, and finally and conclusively as cycles.[29] In keeping with the standard definition of the intellectual outlined earlier, the normal condition, the steady state of the intellectual in any advanced nation, would be disengagement (*dégagement* in French). It is true that for reasons internal to the postwar history of France, the *dégagement* of French intellectuals was somewhat less complete in 1954 than after the cycle had run its course in 1962, or in the United States in the early 1960s, where the intelligentsia was so powerfully influenced by "end-of-ideology" arguments.[30]

Once the undeclared wars began to affect the two intelligentsias—at the end of 1954 in France and at the end of 1964 in the United States—*dégagement* was abandoned. Three stages or levels then followed, each of which can be precisely delineated.

The first stage, which lasted through 1955 in France and through 1965 in the United States, I shall term *pedagogic*. The pedagogic stage was composed of calm, rational, frequently scholarly writings in newspapers and periodicals, in an effort to educate the public and persuade the leaders of the governments in question of the errors of their ways. An important and common ingredient in this level of engagement is the prediction (one might even want to use the word *prophecy*) of what would most likely happen if the governments continued to follow the courses judged wrong or ill advised by the intellectuals. In a extraordinarily large number of cases these predictions were proved correct.

Meetings organized in Paris by the Action Committee Against the Pursuit of the War in North Africa were emblematic of the group aspect of this stage,[31] as was the better-known "teach-in" movement in America, which was pedagogic not only in name but also in origin. The teach-in as a method of conveying information about the Vietnam War and related issues was invented by a professor of anthropology from the University of Michigan, Marshall Sahlins.[32]

The pace of transformation varied among individual intellectuals and the periodicals to which they generally contributed, but by 1956 and 1966, respectively, a significant percentage of the intellectual classes in both countries had moved to new stage, which I shall call *moral*. This does not mean that pedagogic efforts had ceased, far from it, but that they were carried along and incorporated into new dimensions of engagement.

The moral level of engagement involves an ethically based protest and a growing sense of outrage and shame. Jean-François Sirinelli, in his brilliant study of the "war of petitions" that was fought during the Algerian conflict, comments on the motives inspiring those who drafted collective antiwar texts: "Ethics was a motor of their engagement."[33] Feelings of anger and distress were often coupled with a sentiment of confusion and impasse and with uncertainty regarding what form engagement should now adopt.

Another important ingredient in triggering this level of engagement, much more so than is widely believed, was an aggrieved sense of patriotism. In the passion and furor of the times, when it seemed—at least to those holding political power and to conservative intellectuals—that any antiwar protest had to be treasonous, this striking fact was overlooked, whether deliberately or not I cannot say. Examples could be multiplied almost indefinitely.[34] In the French case this motivation has perhaps never been stated more forcefully and with greater clarity than by Pierre-Henri Simon, a reserve officer who had spent five years in a German POW camp, a distinguished Catholic intellectual, a political moderate, and

literary critic for *Le Monde*, later elected to the Académie française. After a particularly brutal FLN massacre, Simon wrote
in 1957 that it had now become even more difficult to alert
the conscience of the French when paratroops or police employed criminal methods in pursuit of pacification in Algeria.
It was natural, Simon argued, to feel "indignation" when an
Algerian rebel assassinated a French soldier or civilian, but
very different was the "intimate suffering" felt at the spectacle
of tortures carried out by Frenchmen themselves: "The former sentiment is the expression of a sociological and visceral
patriotism which is awakened when France suffers, and the
latter is the expression of a patriotism of values and responsibility that is awakened when France sins."[35]

In the American case Mary McCarthy put it beautifully
when she wrote that patriotism had played a large part in her
decision to journey to Saigon in 1967 and to Hanoi in 1968:
"I could not bear to see my country disfigure itself so, when
I might do something to stop it. It had surprised me to find
that I cared enough about America to risk being hit by a U.S.
bomb for its sake."[36]

As he was marching out of Washington to participate in
the famous symbolic encirclement and "siege" of the Pentagon
on October 21, 1967, Norman Mailer (writing in the third
person, as was his wont) recalled that his action "now liberated
some undiscovered patriotism in Mailer so that he felt a sharp,
searing love for his country in this moment and on this day,
crossing some divide in his mind wider than the Potomac, a
love so lacerated he felt as if a marriage were being torn and
children lost."[37]

Again, there are variations of timing, but the third level of
engagement, which I shall term *counter legal*, was largely
reached in France by 1957 and in the United States by 1967.
In both countries significant numbers of leading intellectuals,
often still motivated by patriotism and certainly by moral indignation, began to invoke precedents believed to be estab-

October 21, 1967, The March on the Pentagon. From left to right; Marcus Raskin, Noam Chomsky, Norman Mailer, Robert Lowell, Sidney Lens, Dwight McDonald. [All those pictured are discussed in this book except for the Chicago labor organizer and long-time antiwar activist Sidney Lens. Lens, less of an intellectual than the others, was an important figure in the movement. His contributions to Vietnam War protest are discussed in several works, especially Nancy Zaroulis and Gerald Sullivan, *Who Spoke Up? American Protest against the War in Vietnam, 1963–1975*. Garden City, N.Y.: Doubleday, 1984; and in Lens's own memoir, *Unrepentant Radical: An American Activist's Account of Five Turbulent Decades*. Boston: Beacon Press, 1980.] Photograph by Fred McDarrah.

lished by the Nuremberg trials. (Whether this belief had grounding in international law and historical precedent is not at issue here; the remarkable paralleling of attitudes and the ways that the Nuremberg precedents were evaluated and applied to Algeria and Vietnam will be discussed in later chapters.) Once they had arrived at the third stage, intellectuals were willing to accept the validity of and advocate publicly actions deemed "illegal" by the governments then in power, in an effort to end wars that many had come to view as leading fatally to genocide. The numbers who actually participated in such actions were obviously smaller, though not insignificant in either country especially when we remember that even signing certain petitions placed individuals at risk of government harassment and fines if not actual arrest and imprisonment.

Within this spectrum of counterlegal activities there was wide variation. The majority of the intellectuals who had arrived at Stage 3 insisted on nonviolent approaches—ranging from signing petitions to using class time to talk about the ethics of the wars in defiance of regulations. Other such activities included releasing secret documents, publishing banned books or articles in journals that the editors knew would almost definitely lead to a costly seizure of the press run, refusing to pay a portion of one's taxes, helping draft resisters escape the country, chaining oneself to a strategically located barrier, pouring blood on draft files or even burning them with homemade napalm, and a host of other engagements often remarkable for their ingenuity.

There were serious and sometimes even vicious debates between the engaged intellectuals who believed that civil disobedience was "the last recourse before violence to change a situation which the silent majority has come to look upon as unchangeable"[38] and the active minority that accepted violence and even, at the extreme, aid to forces who were *de facto* enemies of France and the United States—though not *de jure*, as we recall that neither war was ever formally declared. One of my principal contentions is that most French and American antiwar intellectuals, including those who became fully committed to counterlegal activities, remained *engagé*, rather than becoming what I like to term *embrigadé*, that is, abandoning their critical spirit in the unquestioning support of a political cause.[39] This assertion will be tested in Chapters 3 and 4. I recognize, of course, that the boundaries are hazy and that what is true *engagement* for one individual may be *embrigadement* for another and pure folly or even criminality for a third. None of what I shall argue is intended to denigrate *embrigadement*, which in some circumstances may be unavoidable or just or both. I propose that when an individual becomes *embrigadé*, he or she relinquishes the status of intellectual.

Finally, the cycles of engagement drew to a close and at the end of both wars there was a return to what appeared and still appears to be the ordinary life of academics if not all intellectuals, a calm and comfortable existence in the ivory tower. Contemporary observers in France noted the extraordinary rapidity with which intellectuals resumed their normal occupations in the fall of 1962. In the United States full intellectual *dégagement* was not reached until the spring of 1975, after a period of steady deceleration following the Paris accords of January 1973 and the end of the draft.

Stage 1

To illustrate the working of the cycle and to introduce the case studies that follow, I have chosen to begin with examples from our own country, but in which the intellectuals in question refer specifically to the Algerian War. An article by Hans Morgenthau written for the *NYRB* in September 1965 perfectly exemplifies the first stage. Professor Morgenthau, then at the University of Chicago, was an eminent and influential political scientist, a prolific writer, and a political moderate who had served as a consultant with the Defense and State departments and thus had contacts in the Johnson administration. He had initially worked behind the scenes and was able to bring his views on Vietnam to President Johnson's attention. By June 1965 he was willing to go public and so engaged in a celebrated televised debate with presidential adviser McGeorge Bundy. These efforts Morgenthau later viewed as "undertaken in the naive assumption that if power were only made to see the truth, it would follow that lead."[40] Morgenthau was born in 1904 and was, like so many of the finest intellectuals of his generation, an émigré from Hitler's Germany. Thus one might presume that the decision to criticize so sharply the government of the country that had offered him shelter from political violence and terror was especially difficult.

The first part of Morgenthau's article is a careful and professional analysis of the nature of foreign policy, the proper definition of national prestige, the realities of power, and the American misunderstandings of those realities. Then Morgenthau makes an effort at logical persuasion.[41] When, he asks, was the prestige of France higher: "when it fought wars in Indochina and Algeria which it could neither win nor thought it could afford to lose, or after it had liquidated these losing enterprises?" The answer to his rhetorical question is obvious. When France demonstrated the wisdom and courage to end these wars, "its prestige rose to heights it had not attained since the beginning of the Second World War."[42]

Points of transition are frequently illuminating, and later in the same article Morgenthau provides us with a moving illustration of the shift from lucid and rational demonstration to the moral dimension, the second level of engagement. America is involved, Morgenthau writes, in a guerrilla war, and victory in such a war can be accomplished only by genocide. We are following the same path as the Germans did in the Second World War. "We have tortured and killed prisoners; we have embarked on a scorched-earth policy." This indiscriminate killing will get worse, and our armed forces will become brutalized. This is our dirtiest war, and there is no end and no justice in sight. One may reflect, Morgenthau continues, especially if one is worried about national prestige, on "the kind of country America will become when it emerges from so senseless, hopeless, brutal, and brutalizing a war."[43]

Stage 2

By the third year of a full-scale Vietnam war, 1966, a large percentage of the American intelligentsia had followed Morgenthau to the second level of engagement, as had their French compatriots a decade earlier. A powerful example of this stage may be found in Elizabeth Hardwick's almost unbearably poignant "We Are All Murderers," from the March 3, 1966, issue of the *NYRB*.

Ostensibly a review of Jean-Paul Sartre's play, *The Condemned of Altona*, Hardwick's article really addressed the moral questions raised by our presence in Vietnam. Sartre's great political drama, first performed in Paris in 1959, was a perfect vehicle for this message. The play, which superficially is about German guilt during World War II, was in actuality an allegorical representation of French guilt for atrocities committed during the Algerian War.

Hardwick attended the American premiere at Lincoln Center (New York City) and noticed that the audience did not seem to grasp Sartre's message. In the intermission she heard people speaking of the "Condemned of *Altoona*,"[44] and doubted that Americans could "make the leap from Germany to Algeria to ourselves." Hardwick went on to observe that American theatergoers are not used to difficult and complex dramas like Sartre's and that plays that "seek a greater historical and social engagement" have had little success on our stage. Rather, we prefer dramas of "individual neurotic tensions," and one assumes that Hardwick has in mind playwrights like Eugene O'Neill and Tennessee Williams. Maybe, she concludes, our current historical experience is pushing us to "a true meeting with guilt, leading us to suffering, to acquaintance with the sorrows and mysteries and miseries to which *hubris* and power have led other nations." But we are not quite ready yet. None of us—director, actors, audience, and critic, and she does not spare herself—yet "understand what is happening here in 'Altoona' nor what happened some decades ago in Altona."[45]

When the French and American intelligentsias reached the third level of engagement, which they largely had by 1957 and 1967, respectively, they did understand. By 1967 Americans, at least American intellectuals, were growing up. A book like Paul Kennedy's *The Rise and Fall of the Great Powers*, which was a best-seller in 1987, became imaginable. Kennedy links America's destiny with that of other imperialist nations, including France, which have risen to prominence and then retreated to second-rank power status. By 1967 American intellectuals had begun to disagree with a claim made by the editors of *Esprit* in 1959, when they had been in the forefront of opposition to the Algerian War for five years. The editors believed that the crisis in Algeria had become their nation's most fundamental challenge, profoundly questioning France's regime and the values held by French people. "Once again we are confronted with History [their capitalization] more completely and more cruelly than others."[46] This curious reverse chauvinism, which was probably justified in France between 1954 and 1962,[47] was shared by sensitive and thoughtful Americans by 1967.

After 1967 and the Bertrand Russell International War Crimes Tribunal, and especially in the wake of the tumultuous upheavals in Paris in May 1968 which were closely related to and partially triggered by French protest against our Vietnam war, one could legitimately revise the following citation from André Maurois by adding "the United States": "The history of France, a permanent miracle, has the singular privilege of impassioning the peoples of the earth to the point where they all take part in French quarrels."[48]

Stage 3

In the United States we did not fully enter the Counterlegal stage until after the publication in February 1967 in the *NYRB*, and its wide circulation thereafter, of the programmatic essay by Noam Chomsky, "The Responsibility of Intellectuals." This was one of the key documents in the intellectual resistance to the Vietnam War and will be considered in more detail in Chapter 4. There was a flurry of responses following its publication, and in March 1967 the *NYRB* printed a remarkable exchange of letters between Chomsky and the eminent novelist and literary critic George Steiner.

Steiner praised Chomsky for his powerful exposure of the "mendacities that surround us. . . . But what then? You rightly say that we are all responsible, you rightly hint that our future status may be no better than that of acquiescent intellectuals under Nazism, but what action do you urge or suggest?" Steiner wondered whether Chomsky would "help his students escape to Mexico (as Jeanson helped his students leave France during the Algerian Crisis)."[49]

Steiner was referring to Francis Jeanson, a former protégé of Sartre and a philosopher turned activist. During the Algerian War Jeanson was most famous (or infamous from the Algérie française perspective) as the elusive leader of the "suitcase brigades." The exploits of *les porteurs de valises* are thoroughly studied in Hamon and Rotman's work by the same title. Especially when one thinks of the widely proclaimed inefficiency of "egghead" intellectuals, they were amazingly effective in funneling money contributed by Algerian workers across the border into Swiss banks, from whence it was spent on weapons for the Algerian independence forces. Although some of his associates were finally arrested and tried in a widely publicized case in the fall of 1960, Jeanson was never caught by the French secret police, even though he surfaced briefly in Paris in April 1960 for a clandestine press conference attended by a number of journalists.

In October 1957 Jeanson went underground, ten years almost to the day before the publication of the "Call to Resist Illegitimate Authority."[50] He left behind a powerful statement of his own evolution from Stage 2 to Stage 3, which appeared in *Esprit* in May 1957 and should have indicated to the authorities what Jeanson intended to do. His statement was entitled "Para- [i.e., Paratroop] Pacification" and began with an account of the "suicide" of the eminent lawyer Ali Boumendjel,[51] who in a scandalous case "fell" out of a six-story building in Algiers after being savagely tortured by the paratroopers. Yet Jeanson argued that the real drama was being played out in metropolitan France. Governor-General Robert Lacoste was only a sort of henchman and secondarily responsible. "We Frenchmen are primarily responsible, since we consent to this *politique*, in that even in denouncing it we have not managed to erect any obstacles against it. Until there is proof to the contrary this *politique* is ours, these horrors are imputable to us." His friend Boumendjel's case was but one of thousands, and Jeanson believed that France, especially the youth of the nation, was in a perilous state. Rational argument and moral protest had not convinced the leaders of the Fourth Republic to adopt more humane and just policies, and therefore "will not those young people who continue to believe in the values that they were taught in the schools be more and more provoked into taking, against this [immoral] France, the side of its victims?"

In his conclusion Jeanson phrased his question even more explicitly with a direct reference to Vichy France and the Resistance: "For the second time in fifteen years, will official France condemn Frenchmen to betrayal? [*à la trahison?*]."[52]

Ten years later, in March 1967, Noam Chomsky echoed Jeanson in his painfully honest reply to George Steiner's anguished query. He was dissatisfied with the kinds of engagements that I would place in Stages 1 and 2. He had tried harassing members of congress, lobbying in Washington, lecturing, working with student groups, and much more. The only action that Chomsky had taken that had gone further than that of most antiwar intellectuals up to that time was refusing to pay his income tax for the previous two years, a counterlegal engagement in our terminology.

Chomsky was at the time less certain about draft refusal, what the French call *insoumission*, believing that individual decisions should be a matter of conscience. Using language that was similar to that of many French intellectuals during the Algerian War, Chomsky wrote, "I feel uncomfortable about proposing draft refusal publicly, since it is a rather cheap proposal from someone my age." He did advocate tax refusal as a valid gesture, "both because it symbolizes a refusal to make a voluntary contribution to the war machine and also because it indicates a willingness which should, I think, be indicated, to take illegal measures to oppose an indecent government."[53]

Chomsky was well aware "that the limits of possible protest have not been reached." In a widely noted reference to the Spanish civil war, which scandalized some readers, Chomsky pointed out that "thirty years ago many men found it quite possible to join international brigades to fight against the army of their own country."[54] There were, of course, no international brigades established during the Vietnam War, and that kind of engagement (or *embrigadement*) was, and fortunately most would agree, never an option.

Whether during the years of the Vietnam War Chomsky moved beyond Stage 3, out of intellectual engagement and into true *embrigadement*, is a difficult and controversial question. If we take him at his own word, he did not. Chomsky was arrested during the October 1967 march on the Pentagon, as marvelously described in Norman Mailer's *The Armies of the Night*. But neither Chomsky nor the other "invaders" of the Pentagon performed any violence beyond the verbal, against that structure or the massed troops "defending" it.

Chomsky also traveled to Laos and North Vietnam in 1970 as an observer and respected dissident and wrote a powerful account of his trip, published in the *NYRB* and later in book form. But he never advocated violence, and in Chapter 4 I shall provide documentation to show that he was less a victim of "tunnel vision" than is sometimes believed.

In "After Pinkville," an essay that appeared in early 1970 after the Mylai (also known as Song My or "Pinkville") massacre had become publicly known, when the Nuremberg parallels seemed especially convincing, Chomsky rejected violent antiwar protest that would, he thought, be counterproductive and lead to further repression. "Continued mass actions, patient explanation, principled resistance can be boring, depressing. But those who program the B-52 attacks and the 'pacification' exercises are not bored, and as long as they continue in their work, so must we."[55]

In the next two chapters we shall examine that work of which Chomsky speaks so eloquently—the work of two generations of politically involved intellectuals of whom Jeanson and Chomsky are such important members. They were unique in their energy, their courage, their organizing skills, their unquestioned intellectual abilities, and the special varieties of their *engagement*, perhaps their *embrigadement*. Yet they were also representative of much broader movements. We shall analyze in greater detail the work of many of their peers, ranging from Jean-Marie Domenach and the Catholics to Jean-Paul Sartre and the existentialists in France; and from professors near the age of retirement, such as Henry Steel Commager and Hans Morgenthau, to writers who made the trip to Hanoi, such as Susan Sontag and Mary McCarthy, in the United States.

"History," Julien Benda once argued, "is made from shreds of justice that the intellectual has torn from the politician."[56] If we cannot positively identify the shreds, we can analyze the process of the tearing.

3

"This New Dreyfus Affair": French Intellectuals and the Algerian War

The writer's function is not without arduous duties. By definition, he cannot serve today those who make history; he must serve those who are subject to it.... Whatever our personal frailties may be, the nobility of our calling will always be rooted in two commitments difficult to observe: refusal to lie about what we know and resistance to oppression.

"This new Dreyfus affair" refers to the role of intellectuals during the Algerian War,[1] and the following quotation prefaces a "Call to the American Conscience," printed in the February 17, 1966, issue of the *New York Review of Books*. The list of those sponsoring this call reads like a who's who of notable American authors: Eric Bentley, Joseph Heller, Robert Lowell, Norman Mailer, Muriel Rukeyser, William Styron, and Robert Penn Warren, to name a few. Readers of the *NYRB* were invited to join them for a "read-in for peace in Vietnam" at the Town Hall in Manhattan. A chain of "read-ins" stretching from Cambridge to Philadelphia followed in May 1966, with participants including Richmond Lattimore, Anais Nin, Ann Sexton, Susan Sontag, and Richard Wilbur.[2]

There are two links to French intellectuals and Algeria here. Robert Lowell, Norman Mailer, and Richard Wilbur were among the two hundred American intellectuals who added their names to an October 1960 open letter of support for their French colleagues who had risked fines, sanctions, and arrest by signing the "Manifesto of the 121."[3]

Second, the passage just cited, somewhat garbled by the deletion of a long paragraph, was taken from Albert Camus's Nobel Prize acceptance speech, delivered in Stockholm on December 10, 1957, when the battle of Algiers, one of the most violent and bloody periods in "the savage war of peace," had just ended. In the original, Camus speaks of "deux engagements difficiles à maintenir: le refus de mentir sur ce que l'on sait et la résistance à l' oppression."[4]

There is a bitter poignancy in this choice of Camus by a group of America's most talented and thoughtful intellectuals as an inspiration for their own antiwar engagement. When confronted with the war in Algeria and the agonizing decisions it entailed for him as a *pied-noir*, Camus simply could not live up to his own definition of the intellectual's responsibility. This is not to condemn Camus out of hand; indeed, even Amar Ouzegane, a key FLN leader and later cabinet minister in the government of independent Algeria, reflected in 1982 on Camus's "exceptional qualities of lucidity and courage" and praised his openness to the Muslim population.[5] Nor is it to deny that Camus was a writer of uncommon brilliance, eminently deserving of his Nobel Prize. What I shall argue is that the American antiwar intellectuals' reliance on Camus as a model was historically inaccurate, even if effective as a device to recruit additional support for their cause, and that a more appropriate model exists.

Albert Camus: "The Colonizer of Goodwill"

Camus, "the colonizer of goodwill"[6] has been the subject of numerous biographies, and his many glories and occasional failures as a writer and citizen are well known. There is convincing documentation, much of it drawn from Camus's own writings and statements, indicating that in the case of Algeria Camus was unable to sustain the twin engagements that he himself believed essential to the writer's calling.[7] This conclusion is at least partially accepted even by his most ardent defenders. However, the relationship between Camus and two dramatic and painful incidents that occurred at the height of the Algerian War has not been observed. Examining this relationship will set French antiwar engagement in particularly sharp focus.

It is also important to review briefly Camus's own position. Camus collected his principal writings on Algeria and published them in 1958, and they provide an invaluable source. First, it is undeniable that Camus argued forcefully and very early, in his powerful series of articles written in 1939 for *Alger républicain* on the famine in Kabylia, that the suffering of the colonized peoples must be alleviated and that some self-government, what he called an *administrative emancipation*, be granted to the native populations of Algeria, Arabs and Kabyles.[8]

Surely it would not be placing a post-facto interpretation on his words to say that already in 1939 Camus was extremely ambivalent and perhaps even a little guilty about colonization. He wrote in that year that "if the colonial conquest can ever find an excuse, it is in the measure that it aids the conquered peoples to keep their personality." And if we French have a duty toward Kabylia, currently devastated by famine, it is to permit "one of the proudest and most profoundly human populations of this world to remain faithful to itself and to its destiny." A major ingredient of that destiny was, according to Camus, to give lessons of wisdom "to the uneasy conquerors [*conquérants inquiets*] that we are."[9]

In May 1945, after the Sétif massacres, Camus urged the French government not to take violent measures of reprisal, which would be "not only inhumane but also impolitic."[10] But his call was not, as has been noted, heeded. Regrettably, he thought, the desire to become French citizens had dissipated among the Muslim population, because the colonial administration had moved too slowly and because the *grands colons*, had held up assimilation. Camus was already formulating in 1945 what Horne, Talbott, and other historians have called the "lost opportunities thesis."[11] An elemental truth of French policy in Algeria was, Camus believed, that it was "always twenty years behind the real situation."[12] The 1936 Blum–Viollette reform project, which had been abandoned at the time because of intense pressure from the *pied-noir* lobby, would, if enacted, have helped rally the Muslim population to the French state. It was resurrected in 1944, but then it was too little and too late. Camus was convinced in 1945 that in North Africa "nothing that is French will be saved without saving justice." Only through the force of justice will the French be able to "reconquer Algeria and her inhabitants."[13]

After the rebellion broke out in November 1954 Camus became quite pessimistic, even despairing at times, and, taking him at his word, more conservative than he had been in 1939. In 1955 he claimed that Algeria is "our land" and that *pieds-noirs* and Muslims are condemned to live together. Vast reforms are admittedly urgently needed, but "the 'French presence' [*le fait français*] cannot be eliminated from Algeria, and the dream of a sudden disappearance of France is puerile."[14]

As late as 1958, Camus was convinced that the FLN's demand for national independence was "illegitimate." This desire was, he thought, irrational, purely passionate. "There has never yet been an Algerian nation."[15] Camus's strong anti-communism also comes forth here. He believed that the Russians were using Arab nationalism and its Algerian variant for their own strategic reasons. He still could not imagine an independent Algeria, because, in what appears to be a variant of the "domino theory" so prevalent during the Vietnam War, it would lead to "the Kadarization of Europe and to the isolation of the United States."[16] (The cold-war reference may be out of date for some readers: Janos Kadar was the Hungarian politician who after the Soviet invasion in November 1956 and the collapse of the short-lived revolution—the brief period of euphoria and nominal Hungarian independence—ruled his nation until he was ousted from power in 1988. In 1958, the year that former Premier Imre Nagy was executed, Kadar was widely perceived in his native land and in the West as a cowardly leader, a puppet dictator collaborating with the Soviet occupying forces. After 1962 he moved toward the center, and gradually many Hungarians overcame their hatred toward him. Indeed, under his leadership Hungary slowly became more prosperous and tolerant. Kadar died in July 1989, a humiliated and forgotten man.) Albert Camus was wrong about Europe and wrong about the United States.

We shall now turn to a closer analysis of Camus's role, or rather his lack of role, in two crucial incidents that relate to the responsibility of intellectuals and that occurred during the Algerian War. Camus's famous 1957 remark, "I believe in justice, but I shall defend my mother above justice" (*Je crois à la justice, mais je défendrai ma mère avant la justice*), which refracts painfully when viewed in the context of his 1945 statements about justice, raised much controversy and has been widely attacked by writers as diverse as Simone de Beauvoir and François Mauriac and many others. But Camus's defenders have argued, with some justification, that the statement was taken out of context.[17] The statement was made on December 13, 1957, three days after Camus received the Nobel Prize and during an interview with students in Stockholm. It is true that Camus prefaced his comment, which was in response to interruptions by an angry Algerian student militant in the audience, by noting that he had been silent about Algeria for twenty months because the hatred had become so intense that another intervention from an intellectual would only aggravate the terror. But under pressure from his ques-

tioner he would state his views. Camus began with a general
condemnation of terrorism, which at the time was blindly
practiced on the streets of Algiers and "could one day strike
down my mother or my family."[18] His oft-quoted remark
followed.

I am willing to give Camus the benefit of the doubt here
and agree with his defenders that he has been unfairly treated
by his critics. In the same interview, however, there was an
earlier, and I believe more significant, statement by Camus
that has essentially passed unnoticed. Camus affirmed that
despite regrettable press censorship in Algeria, there was a
"total and consoling liberty of the mainland press."[19]

Three months later, in March 1958, the mainland French
authorities seized remaining stocks and banned further publication of *La Question*, Henri Alleg's famous autobiographical
account of his arrest in June 1957 and his subsequent torture
by the French paratroopers involved in the pacification of
Algiers. In his *Les Temps modernes*, when reviewing *La Question*,—which, given the terrible suffering that Alleg endured,
is amazingly free of hatred and the desire for revenge—Sartre
argued that it was the first "optimistic" book to come out of
the Algerian War. It affirmed the possibility of overcoming
apparently insurmountable odds and demonstrated that "this
age of shame and scorn [*le temps du mépris*] contains the promise of victory."[20]

When *La Question* was suppressed in France, the American
publisher George Braziller brought it out almost immediately
in English translation and thought highly enough of it to write
the preface himself. After asserting that this was the first such
book banning in France since the eighteenth century, Braziller
wrote, in language that replicates some of Camus's own (in
The Myth of Sisyphus, for example) that "in an age when in-
difference is the rule, the excitement induced by the implications of one man's account of his trials and his triumph over
them, has given cause for hope."[21]

Henri Alleg had three special links with Camus. He was a Frenchman who lived in Algeria; he was a member of the Parti communiste algérien, as Camus had been between 1935 and 1937; and he had served as editor of *Alger républicain* from 1950 until 1955, when the paper was banned by authorities of the Fourth Republic. Camus had been the principal reporter for *Alger républicain* from its founding in 1938 until January 1940, when it was shut down by the censors of the moribund Third Republic, about to collapse in the chaos of "the strange defeat" of May–June 1940.[22]

In April 1958, after the seizure of *La Question*, its publisher Jérôme Lindon of Les Éditions de Minuit, with the help of the League of the Rights of Man, organized a protest. Four of France's most eminent living writers, who came from extremely divergent political and ideological backgrounds, two of whom were Nobel laureates, agreed to sign a "Solemn Address to the President of the Republic." They were François Mauriac, Jean-Paul Sartre, André Malraux, and Roger Martin du Gard. Martin du Gard, who won the Nobel Prize exactly twenty years before Camus did, had become a personal friend of the younger writer, and in 1955 Camus wrote an elegant, admiring, and sensitive preface to the Pléiade edition of Martin du Gard's complete works.

It was initially planned that because Camus was a Nobel laureate, his name instead of Sartre's would appear on the list. But in the end Camus refused to make this engagement, explaining in a letter to Lindon that even though the "objective was valid," he had decided not to associate himself with any further public campaigns.[23]

The statement that Camus ultimately declined to sign was not overtly political but merely protested the seizure of *La Question*, asked for an impartial public investigation of the charges made by Alleg, and called on the government of the ailing Fourth Republic (the May 13 coup in Algiers that brought it down was only a month away), "in the name of the

Declaration of the Rights of Man and of the Citizen, to condemn unequivocally the use of torture, which brings shame
to the cause that it supposedly serves."[24]

The presence of Camus's name on that list would have
had an important political and moral impact. It would, I believe, have helped lift from the majority of the French people
what Pierre Vidal-Naquet called the "eiderdown of indiffer-
ence" to the practice of torture in this *"sale guerre."*[25] It is to
be regretted that at this point in his life Albert Camus could
not make this simple gesture, which would have fulfilled both
of the engagements he had laid out for the writer just four
months previously.

The second incident we shall examine is the case of Maurice
Audin, which at the time stirred memories of the Dreyfus
affair and retains an exemplary power and pedagogic value.

By December 7, 1957, when Camus left Paris to receive
his Nobel award, the battle of Algiers had been temporarily
won by General Massu's paratroopers, and the city was nominally pacified. As the first groups of draftees who had been
sent to Algeria in 1955 finished their twenty-eight-month term
of service, and even before the publication of *La Question*,
sickening revelations of torture and murder carried out by
the French armed forces in Algeria were coming to the attention of the general public. In his Nobel address on December 10, Camus made only one brief reference to Algeria
but did not mention it by name. Rather, he asked rhetorically
whether he should receive this honor at a moment when his
"native land was experiencing an incessant suffering."[26]

Just eight days before Camus spoke in Stockholm, the
unprecedented defense *in absentia* of Maurice Audin's doc-
toral thesis took place at the Sorbonne. Audin, a young mathematician in the Faculty of Sciences of the University of
Algiers, an antiwar militant and like Alleg a member of the
Algerian Communist party, had been arrested by the paratroopers on June 11, 1957, and disappeared forever. It is now

known that after enduring severe torture he was accidentally murdered by an enraged officer. A false escape was staged; his body was secretly buried and has never been recovered. The case was finally settled in the courts in 1969 but, like the Dreyfus affair, not to the satisfaction of those who wanted full disclosure, who were committed to absolute rather than political justice.[27]

Although attendance at the thesis defense was officially discouraged by René Billères, the minister of national education, more than a thousand people crowded into and outside an amphitheater that could seat only one third that number. By all accounts the ceremony was a moving tribute to Audin, to the highest values promulgated by the French and indeed our own educational systems, and to the very goals outlined in Camus's Nobel Prize address. This was an unforgettable if fleeting moment in the history of education and the history of engagement, when a revitalized "spirit of resistance that would have been shared by a Bernard Lazare, a Zola, and a Péguy"[28] was present in the appeal to conscience and law. Rather than Zola's *J'Accuse* of 1898 the jury's award of the state doctorate with the citation "très honorable" was a "Nous accusons."[29] There was nothing to prevent Camus, who was still in Paris, from attending. But the only Nobel Prize winner whom journalists observed in the audience was the Catholic François Mauriac.[30]

We have reviewed Camus's key ideas concerning colonial Algeria and the Algerian War and have examined two cases when he did not make an engagement, once when he was silent and once when he was absent. What were his engagements? In January 1956 Camus made a courageous effort to bring about a "civilian truce" (i.e., a call for both sides to pledge to halt the killing of civilians). This venture, "a frail bark of tolerance adrift on a sea of fury,"[31] involved a trip to Algiers and considerable personal risk. Although he did not know it at the time, Camus was protected inside the hall where he made his speech, which was located in the Muslim quarter, by clandestine FLN militants, including Cherif Ahmia, a young boxer who later became the champion of France. Outside a crowd of angry *pieds-noirs* shouted "Camus to the gallows!"[32] His appeal was a complete failure, unacceptable to both sides. In outlining his proposal, Camus was not as naive as some of his critics have stated but was aware that a "blind coalition of forces" might be leading to the death of the Algeria he had hoped for with such intensity—a democratic bicultural quasi-independent territory closely linked to metropolitan France. If his hopes were shattered, Camus recognized that he and his small group of like-minded moderates would be obliged, "when confronted with our impotence, to proceed to a total revision of our engagements and of our doctrines, as history for us would have completely changed its meaning."[33]

After the demise of the civilian truce movement in early 1956, Camus lapsed into an anguished and isolated public silence concerning the Algerian conflict. Only rarely did he break his self-imposed rule, in December 1957 when he was abroad in Stockholm and in 1958 when he wrote a brief introduction and a conclusion to the volume of his collected writings on Algeria.

To Camus's great credit, there is unquestioned evidence from a wide variety of sources, including those hostile to him, that in the last three years of his life he was privately and effectively *engagé* in pleading for clemency for Algerian prisoners, obtaining the commutation of the death sentence in a number of cases.[34] Camus was telling the truth when he responded to the young Algerian student who attacked him so bitterly during his news conference in Stockholm, "I can assure you that you have comrades who are alive today thanks to actions of which you are ignorant."[35]

Were it not for his tragic death in January 1960 in an automobile accident, when he was only forty-six, Camus might have decided to make the "total revision" of his engagements of which he had spoken four years earlier. His biographer Patrick McCarthy makes a convincing case that Camus would have broken his silence by the time the Évian agreements were signed in March 1962: "Although one cannot imagine what he would have said, one cannot imagine that he would have looked on as French Algeria and his own past were destroyed."[36]

In 1966, when the "Call to the American Conscience" was issued, Camus had been dead for only six years and was well known and greatly admired in this country. Given the relatively cerebral nature of his writings, the sales of his works are amazing: By 1975, the American edition of *The Stranger* had sold 2.13 million copies.[37] Camus's anticommunism and his particular brand of humanism, which might be termed "Mediterranean secular," had many resonances in American intellectual life in the 1950s and 1960s. Hence it is not surprising that the sponsors of the "read-in for peace in Vietnam" turned to Camus, as he was popularly understood, to sustain their own antiwar engagement.

More perplexing is the case of Father Daniel Berrigan, who, as we have seen, lived in France in the 1950s and knew a great deal about French Catholic radicalism. Yet he, too, on several occasions viewed Camus as worthy of emulation.[38] In the summer of 1968, while awaiting trial for burning, with homemade napalm, draft records in Catonsville, Maryland, Daniel Berrigan posted this quotation from Camus in large letters on the wall of his office at Cornell University: "I wish I would love my country as much as I love justice."[39]

It is one of the paradoxes of our secular age that a far more relevant model—not only to Father Berrigan but also to almost the entire spectrum of the American antiwar movement—would have been provided by the French Catholic intelligentsia, especially but not uniquely those left-leaning intellectuals loosely grouped around the monthly *Esprit*.[40]

"Sin Organized by My Country":
Catholic Antiwar Engagement

The phrase *péché organisé par mon pays* (sin organized by my country) is from an antiwar poem published in *Esprit*.[41] But even in France, *Esprit*, whose circulation stood at 14,000 in 1984,[42] has never reached beyond a minority of the educated elite. Nonetheless, its impact on the dominant classes in French society, the "considerable role" it played in developing a negative attitude toward "colonial realities," has been recognized by Raoul Girardet, probably the most eminent and respected (even by his opponents) intellectual supporter of Algérie française.[43]

Between 1954 and 1962 *Esprit* reflected and documented almost an obsession with what its editors had condemned in November 1955 as "this war without a name."[44] John Talbott entitled his valuable study of the Algerian conflict *The War Without a Name* but does not refer to *Esprit*'s 1955 denunciation. Instead he cites as an epigraph a 1961 statement by Paul Mus, "Le Pays déscend un degré de plus, les yeux fermés, dans une guerre qui ne dit pas son nom." The actual phrase comes from Mus's eloquent and gripping essay introducing a posthumous collection of letters from his son, Lieutenant Émile Mus, who was killed in battle in Algeria in July 1960, eight days before his scheduled release from service.[45] Paul Mus, a world-renowned specialist on Southeast Asia, wrote extensively on Vietnam and was a professor at the Collège de France and later at Yale University. His views on decolonization were cited in Chapter 1.

In one of the many extraordinarily complex and poignant interweavings between Algeria and Vietnam, Mus taught Frances FitzGerald, whose Pulitzer Prize–winning best-seller *Fire in the Lake: The Vietnamese and the Americans in Vietnam* (1972), was extremely influential in finally pushing a majority of educated Americans into an antiwar position. Her book is dedicated to Paul Mus. "I owe most of what I have learned to his wisdom and generosity," wrote FitzGerald in her preface.

Although its editors would obviously claim that no particular honor was involved, it is, I believe, significant that almost five years before Paul Mus's book appeared, *Esprit* could write of "this war without a name." *Esprit*'s involvement with the Algerian War was intense and enduring, with 211 articles published between December 1954 and November 1962, 60 more than appeared in Jean-Paul Sartre's *Les Temps modernes* in the same period. The 42 articles written by Jean-Marie Domenach, codirector of the journal after June 1956 and director following the death of Albert Béguin a year later, would alone fill a sizable volume.

The following analysis will concentrate on *Esprit* as an above-ground journal with a significant readership among France's elite, and the articles and statements of position that the closely knit editorial group determined to make public. Our primary source material will be what was actually published, thus designed for open consumption, after the editorial *équipe* had gone through the sometimes painful process of internal debate and had decided on the public stance to take. The editors were, for example, privately sharply divided over the attitude to adopt toward General de Gaulle after he came to power in 1958.[46] We shall focus on types of engagement that attempted to have an overt impact, rather than the separate, sometimes underground, endeavors of individual *Esprit* group members who were involved in many other activities, as Domenach himself noted in 1987. These activities included the publication of tracts, brochures, and semiclandestine small-circulation papers, particularly *Vérité–Liberté*, whose editor in chief, Paul Thibaud, wrote for *Esprit* and became director of the latter journal upon Domenach's retirement in 1977.[47]

The large volume of materials published in *Esprit* that treat the Algerian War can be roughly grouped into four categories. First, there is the concrete and often perceptive history of the war itself, carefully researched and documented. Many columns were devoted to on-the-scene accounts by native Algerians, French officers and enlisted men, colonial administrators, and *Esprit's* own reporters sent out periodically to review the situation and cover major developments in the war. A great effort was made to be accurate, to verify, for example, any allegations that French troops and police were using torture. The editors believed that it would be dishonorable to take even minor liberties with allegations of this order and were proud that no libel suits were brought against them during the war years. Even their anguish over their perceived inability to affect events, to reduce the documented incidences of tortures, did not drive them to a looseness of vocabulary. Unlike some antiwar intellectuals, who tossed off accusations of "genocide" for something as innocuous as the Constantine Plan for the economic development of Algeria, the editors of *Esprit* would not use this highly charged word in denouncing the criminality of the French state in Algeria.[48] (Indeed a semantic, perhaps even a semiotic, analysis could be made of the ways that intellectuals, during both the Algerian and Vietnam wars, employed the terrible word *genocide*, which after all normally means the premeditated and methodical extermination of a people.)

Besides *Esprit's* attempt to be as accurate as humanly possible, another exemplary element in its running history of the Algerian War was its balance and fairness. *Esprit* opened its pages to a wide variety of commentators on the war, including a number of non-Catholics. Never did it follow a single political "line," with documentation selected to suit that line. The *Esprit* group was, to be sure, unanimous in its opposition to the war, *engagé*, but it never adopted the tunnel vision of the *embrigadé* militant.

On many occasions the *Esprit* editors denounced the brutality of the FLN forces, and the generous *tiersmondisme* that many of them shared did not blind them to iniquities present in Nasser's Egypt.[49] In 1960 the *Esprit* group announced that its members would protest strongly if a right-wing leader like the detested Jacques Soustelle, the governor-general of Algeria in 1955 and 1956, who had a primary responsibility for establishing the concentration camps there, were placed in one himself by the Gaullist regime.[50] Soustelle, a man of unquestioned brilliance whose qualifications as an intellectual would be difficult to dispute, was a distinguished anthropologist and respected leader of the anti-Nazi Resistance. Like many who came before him, the natural beauty and particular ambience of Algeria went to his head—*Algérie montait à la tête* was the phrase often used. Soustelle abandoned his liberalism in 1955 and moved 180 degrees politically, under circumstances that have never been fully elucidated. He became minister of information in de Gaulle's first government in 1958, was a shrewd and effective supporter of Algérie française, and barely survived an FLN assassination attempt. In 1959 he broke with de Gaulle as it became clear that the general was moving in the direction of granting independence to Algeria.

Toward the end of the war *Esprit* went as far as to condemn and publish evidence of the use of torture by the French police against members of the OAS, a secret army organization.[51] In a last-ditch struggle to keep Algérie française, the OAS launched a reign of terror beginning in 1961 in Algeria and mainland France. It was feared for its brutality and infamous for its use of plastic explosives against civilians.

The second broad area of *Esprit*'s concern with the Algerian War was a careful and frequently astute study of French internal political life between 1954 and 1962, as it reflected and was often dominated by the conflict across the Mediterranean. Close scrutiny was, of course, paid to the change of regimes in May 1958, triggered by the military plot in Algiers.

*

Third, one finds a fascinating history of antiwar engagement,
focusing on but by no means limited to Catholic activities.
Attention is given to Protestant involvement, for example,[52]
and there is a sensitive and open-minded evaluation of the
actions of Francis Jeanson who had contributed occasionally
to *Esprit* before he went underground in 1957.[53] The issues
relating to conscientious objection are addressed with great
sensitivity and balance. Americans who counseled conscientious objectors during the Vietnam War, as I did, will find
that these articles provide a haunting sense of *déjà vu*. The
thorny question of selective conscientious objection, which
caused major intellectual, moral, and political problems for
American draft counselors, is examined, along with the issue
of alternative civilian service.[54] There are accounts of sit-down
techniques, similar to those used by draft resisters during
Vietnam, and a report of an incident in 1960 in which four-
teen persons were arrested, all claiming to be the same draft
evader (another common Vietnam-era method). Like the crip-
pled Vietnam veteran in the famous film *Coming Home*, they
chained themselves to metal grillwork, in this case that sur-
rounding the Jardin de Cluny.[55] And there are reports of
trials of conscientious objectors that could, with a few names
and places changed, have been set in American courtrooms
exactly a decade later.[56] To resurrect a metaphor used in
Chapter 1, the xerox machine here was copying quite perfectly.

Fourth and finally, there was in *Esprit* between 1954 and
1962—using as a basis for reflection the varieties of antiwar
engagement it documents in its pages—a serious and ongoing
meditation on the Catholic intellectual's responsibility to contemporary society and what forms his or her engagement
should take. This meditation was more systematic than that
found in, for example, the *NYRB* (or even *Ramparts* or *Commonweal*, journals that at least began with a liberal Catholic

perspective), during the Vietnam War years. *Esprit*'s reflections on Catholic intellectual engagement may easily be generalized to all members of the intellectual class; they closely follow the evolution of the war in Algeria, drawing from it what might be termed a *negative inspiration*. One is not surprised to discover these concerns in *Esprit*. In this journal, immediately upon its founding by Emmanuel Mounier in 1932, the very conception of "engagement," in its modern sense of the political involvement of intellectuals, was articulated and popularized.[57] To elucidate the particularly sensitive and acute way that *Esprit* reflects the cycle of engagement, the following analysis concentrates on this fourth area, though where appropriate we shall draw from the other three categories.

Stage 1

Fittingly enough, the first of the more than two hundred *Esprit* articles on the Algerian War was drafted by Jean-Marie Domenach and appeared in the December 1954 issue. Given the time delays in the publication of a monthly, it must have been written shortly after the rebellion began on November 1, 1954.

In this almost eerily prescient first article, suggestively entitled "Is It War in North Africa?"[58] Domenach makes no call for formal engagement but implies that at this juncture the role of the intellectual should be limited to intelligent reporting and the offering of warnings and advice to those in authority. Hence we are squarely in Stage 1, what I have termed *pedagogic*. Already Domenach was unsure whether France would have an easy victory over what at the time appeared to most observers, including France's political establishment and the prime minister, Pierre Mendés-France, to be an inexcusable, indeed criminal, and minimal outbreak of violence in the countryside, led by a tiny group of fanatics with no real support among the Muslim population. Domenach, on the other hand, wonders whether his nation will face a bitter war similar to the one that has just ended in Indochina. He finds it striking that his compatriots have not wished to see that in the previous twenty years Algeria has become a nation, a point that Albert Camus could not comprehend as late as 1958, and that when one prevents people from voting, speaking, and writing as they wish, they end in armed rebellion. Domenach demands strict disciplinary measures against the police who in some Algerian cities have "had recourse to torture."[59] He concludes by calling for a wide spectrum of economic and social reforms to bring justice and an equitable standard of living to the Arab population. If these measures are not quickly carried out, "it will be necessary to bring in the paratroops, the CRS [Compagnies républicaines de sécurité, the elite special police], the regular police, and the full complement of draftees, to sustain an interminable war."[60] All of his predictions proved to be perfectly accurate.

The same basic position—informational, reformist, historical—continues for almost a year, with a growing tone of anguish and frustration, indignation and embarrassment at the crimes being committed in the name of France. By July 1955 *Esprit* could print an article by Colette Jeanson, who had just returned from a trip to Algeria. Jeanson claimed that a full-scale war was already under way there, with all its accoutrements of internment camps and displaced persons. She hints at the form of engagement that she and her husband eventually adopted when she writes that "the 'outlaws' are not those whom one would think."[61]

Esprit welcomed a remarkable contribution by François Sarrazin, submitted from Algiers in July 1955. Sarrazin was the pseudonym for Major Vincent Monteil, an officer fluent in Arabic who was Soustelle's military adviser until he could no longer tolerate the direction that French policy was taking and so left Algeria. He later converted to Islam. Given his position he knew firsthand that Algeria was a "country without law." Sarrazin reviewed the chain of broken promises dating back almost to the conquest of 1830, down to the governor-general's action of 1948, taking from the Algerians through rigged elections "all legal possibility of self-expression." Seven years before the mass exodus of *pieds-noirs* Sarrazin was already aware that it might be too late to save a French presence "worthy of the name" in Algeria. He believed, again correctly, that the Berber (Kabyle) population was sufficiently Arabized to support the FLN. We are observing, Sarrazin wrote, "the birth of a nation." We French have had 125 years to assimilate Algeria, and it is time to be honest and recognize our failure. In his own variation of the "lost opportunities thesis," Sarrazin argued, again with firsthand knowledge, that the colonial authorities were always one reform too late—*toujours en retard d'une réforme*. He understood long before his political and military superiors did that "Algeria *is not* France."[62]

In a second and equally striking article, published only a month later in *Esprit*, Sarrazin emphasized moral factors more strongly, noting that the torture carried out by the French police was "worthy of the Gestapo." He feared that Algeria was close to a real race war and noted that a probable result of the brutal French repression would be the forging of a nation in clandestinity. The tortured prisoners would be the heroes and martyrs of tomorrow. In another accurate forecast Sarrazin predicted that one day their names would be given to the streets of their native city.[63]

Stage 2

By the end of 1955 *Esprit* was moving, with some hesitation and uncertainty, to the second level of engagement. The language is stronger, and the moral tone more and more pronounced. The entire November 1955 issue, dedicated to the Algerian question, was entitled "Let Us Stop the War in Algeria." The *Esprit* group now spoke of a "falsely French Algeria," and although they still expressed some hope for friendship and contact with Algerians beyond the "abominations of this war without a name," they believed that if radical reforms were not soon undertaken, the French government would find itself in a war that the French populace could not recognize as its own. What remained of national cohesion would probably disappear, but the editors of *Esprit* insisted that they would "never become accomplices to a violence against which we were in the past irreversibly engaged" [*une violence contre laquelle nous nous sommes engagés naguère sans esprit de retour*].[64] The past violence to which they are referring is that of Nazism. The elder generation of the *Esprit* group, those who reached adulthood during the Second World War, have always claimed for themselves impeccable antifascist and Resistance credentials.[65]

The position that *Esprit* adopted at the end of 1955 involved a kind of distancing from the government and its policies, but one that was still private and primarily intellectual. Like Simone de Beauvoir, François Mauriac, Henri Marrou, and so many others and like so many of their *confrères* in the United States a decade later, the group of writers associated with *Esprit* suffered a painful and growing sense of shame at their French nationality.[66]

The strong focus on the Catholic and thus the moral ingredient in their engagement remained. Domenach was proud of the way his church's attitude had changed from its long-standing conservatism. The French church, as is well known, had almost universally been on the side of the anti-Dreyfusards and more recently had ignored the massacres in 1945 in Sétif and those in 1947 in Madagascar. But at the end of 1955 both on the mainland and in Algeria the church had taken a stand in a "firmly spiritual style." For example, the bishops of Algeria, especially their leader, Monseigneur Léon Étienne Duval, the archbishop of Algiers, were calling in 1955 for an end to reprisals and for reforms to improve the miserable standard of living of the Muslim Algerians. As a result Duval was bitterly attacked by the *pieds-noirs*; children called him "Mohammed" in the streets, and "worshipers" constantly interrupted his services in the Cathedral of Algiers.[67]

Raoul Girardet is correct that within French Catholicism there is not a single name of an antiwar intellectual, a single attitude, a single form of engagement, "against which one could not contrast in parallel, in the adverse camp, other names, other attitudes, other types of engagement." The paratroops had a chaplain, for example, who found excuses for the use of torture, and numerous clergy supported Algérie française. Just as the new revolutionary current in American Catholicism represented by Father Daniel Berrigan was counterbalanced and indeed overweighted by Cardinal Francis Spellman, an ardent supporter of the American war effort in Vietnam who made regular visits there to bless the forces, so too could Monseigneur Jean Rodhain, a leading conservative figure in the French church, make a statement in 1960 remarkably similar to "America: Love it or leave it" of a decade later. In denouncing conscientious objection, Rodhain asserted: Either serve in the armed forces, "or if you really do not want to fight, then go and live in another country."[68]

But Domenach is equally correct that the powerful segment of French Catholic opinion that had always been on the side of the "maintenance of order" could no longer claim to be the sole representative of the church.[69]

To pursue a bit further the question of the attitudes of the French Catholic hierarchy toward the Algerian War: Recent research has shown that Monseigneur Duval came out unequivocally against torture on January 17, 1955, when the war had been under way for only two and a half months. In October 1956, in a circular to his clergy that has never been published, he stated that in the establishment of new relations between Algeria and France, it will be necessary to take into account the progressive satisfaction of the will for the "*autodetermination* of the populations, in the respect of the rights of individuals and communities."[70] The term *autodetermination* is normally believed to have been one of the several brilliant neologisms created by General de Gaulle, this particular one in a declaration of September 16, 1959, as he gradually pushed the French people and/or was propelled himself toward accepting the necessity and inevitability of Algerian independence.

As the undeclared war dragged on and the level of violence escalated during 1956, *Esprit*'s profound concern with the issues raised by the conflict never wavered. In *Esprit*'s many pages on Algeria we continue to find the pedagogic ingredient: a combination of astute political analysis, careful reportage, and pleas for a peaceful settlement before it is too late for any French presence to remain in Algeria.[71] But the pedagogic approach was now coupled with the moral appeal to Christian conscience, as reiterated by Domenach in February 1956. He reminds his readers that it is not necessary for a Christian to opt for a particular political formula, such as federation or integration, as a pragmatic solution to the Algerian problem. But to sit passively by as accounts of torture flow in, claiming that one can do nothing to help the Algerian people or to end the war with justice, is in fact taking a temporal position. The attitude of abstention, which Paul Nizan had already pointed out in 1932 is a form of choice,[72] represented for Domenach "the abdication of the spirit when confronted with the supremacy of force."[73]

On the military and political fronts, one of the contributors to *Esprit*, Jean Sénac, in March 1956 predicted an Algerian victory,[74] and in April of that year Alain Berger warned with equal prescience that the Fourth Republic was in danger of a rightist coup.[75] Some of *Esprit*'s attention at this time accordingly shifted away from an attack on the war in Algeria to a defense of the republic at home.

As it was for many communist intellectuals who resigned from the party in protest, and as it was for Jean-Paul Sartre whose ideological trajectory veered dramatically,[76] the brutal suppression of the Hungarian revolution by Moscow in November 1956 was a severe setback for *Esprit*. The December issue was devoted to "the Flames of Budapest." However, for Domenach, in a piece entitled "Our Fault," the crimes of Stalin and his successors did not absolve the French for what they had done and were doing in Algeria. Those ethical dilemmas would not magically disappear as the result of a simpleminded moral mathematics of equivalency. The war continued to rage, and the French army and police were employing the same methods. Independent Catholic and other leftist intellectuals would simply have to correct "the form and the method of our engagement."[77] Collaboration with the French communists would be impossible until they disassociated themselves from the counterrevolutionary terror in Hungary.

Stage 3

In January 1957 *Esprit* moved toward the counterlegal stage. Jacques Julliard, who has become one of France's most eminent historians and was at the time a leader of the national student organization (the Union Nationale des Étudiants Français, or UNEF), joined with Domenach in drafting an important programmatic article. They called for decolonization, economic modernization, and popular education. It was no longer adequate for a writer to be a specialist in "moral protest," as even virtuous indignation is unhealthy "when it does not include any engagement."[78] The clear implication of this statement is that prior antiwar activities no longer deserve to be portrayed as engagement. Domenach and Julliard were still convinced that the form of their engagement must, in a strict sense, remain intellectual, that is, rely on the printed page. Yet this had come to mean facing arrest, fines, and costly seizures of issues of *Esprit* by the government. Because propaganda consists of changing the name of things, it had become a "political crime to point out errors in vocabulary."[79]

There was no rhetorical exaggeration in this statement. In the end pages of *Esprit* throughout these years there is a leitmotif of reports of seizures of the journal, arrests of group members, and pleas for extra funds from their loyal readership to cover the resulting financial losses. The headquarters of *Esprit* was also bombed twice with plastic explosives by the OAS, and many records were destroyed.

Early in 1957 the editors of *Esprit* refrained from publishing documented accounts of the use of torture by French police and military in Algeria, under the assumption, soon proven erroneous, that holding back materials that would humiliate conservative or misguidedly patriotic Frenchmen would help peace negotiations.[80]

By May of that year Domenach returned to the theme of moving beyond indignation. Moral indignation does not really challenge the system that makes such crimes possible. It is highly significant, and more than a coincidence, that just as *Esprit* was struggling with how to go beyond the second level of engagement we find the first reference in its pages to the Nuremberg parallels. The two themes are intimately related here and elsewhere, because most if not all practicing intellectuals after 1945 possess a historical consciousness.

At the Nuremberg trials, Domenach pointed out, ordinary German soldiers did not stand among the accused but, rather, the leaders who ordered the common soldiers to make war. In the dock at Nuremberg in 1946, and again in the Fourth Republic of 1957, the guilty ones were and are the high-ranking officers, along with the politicians who hide the existence of war crimes and simultaneously, if illogically, try to habituate the populace to them. Confronted with this uncomfortable reality, concerned French intellectuals must have as their goal more than the moral aim of salving consciences, they must struggle to bring about peace.[81]

Much uncertainty remained regarding a specific course of action that could help reach this end. One approach was indicated by Francis Jeanson who, also in the May 1957 issue of *Esprit*, made his last contribution to that journal for the

duration of the Algerian War. Jeanson openly called for counterlegal action,[82] the precise outlines of which were as yet
undefined, as he was just moving into clandestinity.

For all the esteem and friendship its editor had for Jeanson,[83] *Esprit* was still searching for a proper path, trying to
find a just and justifiable and not totally useless and frivolous
form of engagement. That quest continued through September 1957, when Yves Goussault, a new contributor, published
the first report to appear in *Esprit* on a group demonstration
for peace. It was a silent protest, which after having been
planned and canceled several times finally took place in June
1957, in Paris in the gardens of the Palais Royal, with only
five to six hundred present. Although it was banned by the
government, it was not bothered by the police until after the
protesters moved out of the park and headed down the Avenue de l'Opéra. Forty-nine were then arrested. Goussault
viewed this action as a new form of protest, "inspired by the
Anglo-Saxons."[84] Goussault did not indicate which Anglo-
Saxons he had in mind. That year, 1957, was still too early
for significant ban-the-bomb activity in England, and in the
United States the civil rights movement was just getting under
way. In any case Goussault's intriguing observation indicates
from the French side the complex nature of the cross-
fertilization involved in contemporary intellectual engagement.

This demonstration was, according to Goussault, a bearing
of witness (*témoignage*) by both atheists and believers. Jean-
Paul Sartre and François Mauriac joined in the march. Although Sartre's presence was not surprising, Mauriac was
seventy-two years old at the time and besides being a Nobel
laureate was one of the forty "immortals," a member of the
Académie française, a staid and heavily pro-Algérie française
group. "Mauriac in the streets! That is not insignificant." Still,
Goussault was uncertain of the utility and validity of this demonstration when the situation was so desperate. "Never have

June 23, 1957, François Mauriac and Jean-Paul Sartre immediately after the silent demonstration protesting the war in Algeria. While the two intellectuals do not look particularly happy to be in each other's company, it is significant that they temporarily put their vast ideological and philosophical differences aside and made the commitment to march together. [No photograph is known to exist of the actual demonstration, which was banned by the police. In his elegant *Bloc-Note* describing the event, published in *l'Express*, June 28, 1957, Mauriac explained that since the police were present in large numbers in the Tuileries, the demonstrators assembled in the interior gardens of the nearby Palais-Royal, where their silent march was observed by only a single individual.] Photograph courtesy of Agence France-Presse.

we Catholics sensed so profoundly both the absurdity of this war and our own impotence, and we must be looking for "new forms of action."[85]

The Ethic of Distress

The situation was clarified and the new forms of action delineated in March 1958 in a brilliant and influential article by the Protestant philosopher Paul Ricoeur, an occasional contributor to *Esprit*. In this article Ricoeur advocates counterlegal engagement, the betrayal of official legality. He uses as a pretext for his discussion the case of the Protestant pastor Etienne Mathiot. Mathiot had been arrested in Besançon for sheltering an FLN leader and guiding him across the Swiss frontier. Mathiot defended his action by stating that his guest was a political official and not an assassin and that he wanted to spare the man's being tortured.

Ricoeur takes this opportunity to present a forceful validation of counterlegal engagement involving the concept of the "ethic of distress," which comes into play only under certain historical conditions. When a nation sinks to the level of illegal violence, even if some kind of enabling act has been rammed through a legislature to make the violence in question spuriously legal, the nonviolent gesture of a citizen like Mathiot, although illegal in a narrow sense, is acceptable.[86] In defending Mathiot's engagement, Ricoeur also plays on the distinction between the "nation" and the "state" so eloquently raised in the conclusion of Robert Paxton's classic study of Vichy France, in which Paxton observes that "there come cruel times when to save a nation's deepest values one must disobey the state."[87] For Ricoeur such disobedience in 1958 is the product of the demoralization of the nation by the state. He goes on to draw the religious parallel by referring to the temptation of Pontius Pilate, which Pastor Mathiot wished to exorcise by his deed. Mathiot's act was not a "sterile gesture of protest" but has validity as an affirmation of Christian eth-

ics, and in the future it may have the positive political effect
of reminding Islamic Algerians that friendship with France
is still possible.[88]

Thus, under the pressure of events, as the Fourth Republic
was about to collapse as the result of violent illegal activity
centered in army units based in Algeria, *Esprit* came to the
conclusion that the nonviolent breaking of the law of the land
was a justifiable form of engagement. Its members would not
then and never did in the remaining four years of the Algerian
War endorse violent action; they could not agree with the
methods of Jeanson's suitcase brigade. Rather, their position
was similar to that adopted by Father Daniel Berrigan.
Whereas Berrigan went into hiding for several months and
was almost as embarrassing to the FBI as Jeanson had been
to the French equivalent, the DST, Berrigan strongly denounced the violence of the Weathermen even as he sympathized with their anger and alienation.[89]

In discussing the related issues of conscientious objection
and draft resistance, none of the contributors to *Esprit* ever
advocated actual desertion.[90] They would support young men
who had already refused induction or gone AWOL and would
offer them refuge. But as Domenach argued in December
1959, for intellectuals to encourage such disobedience would
be irresponsible, even "vile."[91] They would face minimal pen-
alties, at the most a fine, and rather than attempt to persuade
soldiers on active duty to risk imprisonment or the firing
squad by deserting, they should work for peace in Algeria.

This attitude was almost exactly replicated by the core of
the draft-counseling movement in America during the Vietnam War. In general, draft counselors advocated and practiced what was called "informational" rather than "political"
counseling. Counselors, most of whom were intellectuals,
would present all the available options to their clients—including claiming conscientious objector status, facing trial with
its probable result of imprisonment, and evasion to Canada—

rather than attempting to pressure them to refuse induction or to desert if they were already in the army.[92]

One further important clarification of *Esprit*'s position on counterlegal engagement occurred in April 1960, after the dramatic arrests in February of more than twenty metropolitan Frenchmen and women, members of the Jeanson network accused of supporting the FLN (among those indicted were actors, students, and a priest). The charges against them were divided into three categories:

Aiding draftees who refused to serve in Algeria to escape across the French border.

Sheltering Algerians sought by the police.

Diffusing propaganda and performing other services for the FLN.

Domenach believed that the first two categories of actions could derive from Ricoeur's "ethic of distress," an ethic that permits recourse to nonviolence to "remedy unjustifiable violences."[93] Under certain conditions it gives one the right and perhaps the duty to resist the repression of a government, even if one goes against written law. Domenach acknowledged that in the specific situation of France in 1960, actions impelled by the ethic of distress might indirectly help the FLN's cause. However, direct assistance to the FLN was of a different nature, and *Esprit* could not approve it. (They do not use the term *embrigadement* to describe such assistance, and there is no empirically satisfying way to prove that it was. I would argue that the contextual analysis that Domenach uses to defend *Esprit*'s position makes an extremely persuasive case that Jeanson and his associates were indeed *embrigadé*.)

The *Esprit* group was convinced that the prosperous and relatively free France of 1960 was simply not at a moment in its history when disaffected intellectuals should break totally from their country, disavow their government, and combat its (their own) army. In the immediate past such moments had existed, and Domenach admitted certain parallels with the France of Vichy, such as the Gestapo methods used by the forces of order in Algeria. Nevertheless, the situation had not gone far enough for the intellectual to forgo all allegiance to France. After all, the Fourth Republic had decolonized in Tunisia and Morocco, and the Fifth in black Africa. The intellectuals' goal should still be to struggle for a negotiated peace in Algeria, not for the victory of the opposing camp.[94]

Domenach would not allow himself to be pushed into clandestinity. De Gaulle was no more of a tyrant in the spring of 1960 than he had been a year earlier, and the Fifth Republic was decidedly a "soft tyranny."[95] How could de Gaulle be blamed for taking over the state in May 1958, when "it was given to him"? Obviously de Gaulle encouraged the donation, but the statement is hard to challenge, even with the benefit of historical hindsight.[96] Domenach believed that the responsibility of intellectuals was to demonstrate that "between the frivolous word and the recourse to arms there exists a path."[97] The path should be one of nonviolent resistance, civil disobedience, and peaceful protests. Teachers should work to educate their students, breaking regulations by reading antiwar texts during class, for example. Before asking people to risk their lives, Domenach suggests, we might advise them to risk their jobs. We must awaken the dormant consciences of those who had decided to leave the settlement of the Algerian crisis to Charles de Gaulle, known to all Frenchmen and women simply as *mongénéral*, their paternalistic, arrogant, sphinxlike, but certainly not fascist president.

The path that Domenach urged his fellow intellectuals to follow in May 1960 thus contained a variety of engagements, both counterlegal and legal, both individual and group. In making this range of proposals, Domenach's aim was to deliver the Left from its paralyzing sentiment of impotence, to act for all those who believe there are better things to do than "cultivate a despair that is the secret weapon of tyranny."[98]

Despite nearly giving way to despair after the failure of the Melun negotiations in the fall of 1960, which they totally blamed on French intransigence and which led Domenach to waver momentarily in his nonviolent stance,[99] the editors and contributors to *Esprit* remained faithful to these positions.

None of the then current editorial staff, and only one individual associated with *Esprit*, François Wahl, signed the Declaration on the Right of Insubordination in the Algerian War (the Manifesto of the 121), which circulated in the fall of 1960, eventually receiving 246 signatures.[100] In the wake of the scandal caused by this manifesto, Domenach and Ricoeur joined union and student leaders, other eminent intellectuals such as Roland Barthes, Georges Canguilhem, Maurice Merleau-Ponty, and the historians Ernest Labrousse and Jacques Le Goff, in signing a more temperate appeal, "For a Negotiated Peace in Algeria."[101] The *Esprit* group did assert that they would go to prison with Jean-Paul Sartre and the 120 other original signers of the more famous and more inflammatory manifesto if the government took punitive legal action against them.[102]

This disagreement over wording and the resultant multiplication of manifestos and duplication of effort illustrate how difficult it was to accomplish Domenach's stated goal of mobilizing and unifying the Left around the theme of opposition to the Algerian War. One of the ideas put forward in Domenach's May 1960 appeal was a call for a truly mass demonstration, which had not yet occurred in France, to put pressure on the government to negotiate a peace settlement. In large measure because the French Communist party never lent its support, it was not possible to organize something similar to the "spring mobilization" of April 15, 1967, in Manhattan, when somewhere between a quarter of a million and a million demonstrators (depending on who was counting; crowd estimating remains an uncertain science) marched from Central Park to the United Nations Plaza.

Although it was a dramatic increase over the few hundred

who gathered in the Tuileries in the summer of 1957, Domenach reported with great regret that as late as December
19, 1961, only fifty thousand out of a population of seven
million in the Paris region came out for a march from the
Bastille to the Hôtel de Ville to protest OAS terrorism. Even
though the Gaullist government was locked in a struggle with
the OAS at the time, the demonstration was banned, with
Minister of Information Louis Terrenoire asserting, completely fallaciously, that it had been organized by the Communists. The demonstration was broken up by the police and
could never fully deploy. More than a hundred marchers were
hospitalized as a result of police clubbing. Domenach reported
angrily that "the government is turning against those who
wish to support it."[103]

A gigantic, nonviolent, orderly, and authorized peace
march finally did take place on February 13, 1962, when a
crowd conservatively estimated at one-half million demonstrated in Paris, a month before the Évian agreements were
signed. This massive protest may have slightly hastened the
peace process, though negotiations were already well under
way and approaching closure. Tragically the French Left
ceased squabbling and organized this huge march only after
eight ethnic French demonstrators had been killed five days
earlier at the Charonne métro station, when they were trap-
ped and suffocated by a police charge during a much smaller
peace march, which had not received government authorization.[104]

One final modification of *Esprit*'s position on intellectual
engagement was made in November 1961, after the savage
repression of the first demonstration on French soil by ethnic
Algerians. On October 17, approximately seventy thousand
Algerian workers, dressed in their Sunday best, had marched
peacefully and unarmed through the streets of Paris, protesting a curfew, and police brutality was extreme. A conservative estimate is that twelve thousand were arrested and held
for several days in special camps that were barred to journalists and priests; a thousand were deported to Algeria; and

no one knows how many were secretly liquidated. This highly visible exportation to the mainland of police methods that had been more or less openly employed in Algeria since 1954 struck a raw nerve, and *Esprit* responded by offering its first collective instruction (*consigne*). Previously the usual approach had been to portray situations demanding ethical action and leave it to the individual to decide whether he or she wished to become involved. Now *Esprit* openly asked its readers to demonstrate in groups, to oppose racism, to alert civil and spiritual authorities, to form associations, to multiply protests, and incessantly to call for peace in Algeria.[105]

Closing the Cycle

This remained *Esprit*'s stance with regard to engagement and the responsibility of intellectuals until the peace negotiations finally came to fruition five months later. At that time, *Esprit* published a number of sensitive and thoughtful retrospective articles, which are notable for their lack of recrimination.[106]

In only one of the articles was the tone bitter. In "*Paris-Match* at the Hour of the Cease-Fire," Philippe Ivernel first offered a subtle and witty analysis of the "radiant lies" (*mensonges radieuses*) about the reality of the Algerian War promulgated by this mass-circulation weekly. Through a whole series of photographs and captions *Paris-Match* almost pretended that the war had never taken place. The FLN militants and the soldiers in the new Algerian army, the ALN, which had been largely trained and organized in camps in Tunisia and which during the war *Paris-Match* had portrayed as rebels armed with "knives, hatchets, and iron bars" were now shown marching in order in neat uniforms. It all was sanitized; the brutality, the horror, the 800,000 deaths were ignored.

This was Ivernel's estimate in the immediate aftermath of the war. "Who will ever know the exact number?" he admitted, and his guess is not as far off the mark as some. The FLN claimed 1.5 million; Jean-Paul Sartre, writing in September 1961, asserted that there had already been a million Algerians killed. The French military claims only 141,000. The figure of 600,000 dead, counting Algerian and European casualties, is a more probable total that is accepted by scholars.[107]

By the spring of 1962 the government had managed to break the power of the OAS in mainland France and capture its ringleaders, although in Algeria it continued its savage last-ditch struggle until concluding a truce with the FLN in June, a month before formal Algerian independence. In *Paris-Match* all we see of the extremely violent OAS is General Salan and the other arrested officers (who had led a putsch against the government in April 1961) comfortably incarcerated in the Santé prison, wearing elegant suits and dining at a well-laid table. The next photograph shows them happily playing volleyball in their prison courtyard. The war, according to this media presentation, was mere smoke, a bad dream, and will leave behind hardly more than a memory.[108]

Ivernel concludes by referring to an exchange of correspondence published in *Le Monde* just after the cease-fire, in which important bureaucrats who had served in Algeria excused themselves and one another. What happened was a sort of mass-produced redemption (*une redemption en chaine*), and French officialdom blamed itself "not for vile criminality but for insufficient grandeur of soul." Nevertheless, argues Ivernel, we still need to understand, as the Algerians have agreed not to call us before the bar at Nuremberg, how our country, "full of noble sentiments and decorated with such a good conscience, was able to give birth to a monster, the war in Algeria."[109]

Surely *Esprit* is one of the most helpful available tools in understanding that birthing, that *accouchement*, and surely there is neither exaggerated bitterness nor cynicism in Ivernel's portrayal of what he views as the indecent haste of his compatriots to put the Algerian War behind them. Anyone who lived in France during and after the Algerian War would agree with the final words of John Talbott's excellent history: "On the subject of Algeria a great silence fell over the land."[110]

There is a wide consensus among scholars from a range of disciplines who have studied this period that the Algerian War was masking a number of French social, cultural, and economic realities—those connected with the postmodern era, the age of consumption, technocracy, the Concorde and the Common Market, mass media, the ethic of vacations, Club Méditerranée, and all the rest—and that needed only the settlement of the Algerian crisis to appear in full force. In finally terminating its last colonial war France was ready to "marry her century," in de Gaulle's terms, if not quite in the more heroic and less introverted way that the general had anticipated or at least hoped. Alain Touraine, one of France's leading sociologists, makes this point beautifully in asserting that the Algerian War prevented his colleagues, and he includes himself among them, from possessing a clear vision of society. We were locked into old images: "After the peace in Algeria, France changed in a week."[111]

In an intriguing way both *Esprit* and General de Gaulle were in agreement here, and both were to be somewhat disappointed. The September 1958 issue of *Esprit* had been entirely dedicated to an analysis of the new Fifth Republic, and the editors made an interesting admission in a prefatory note. They reminded their French readers that on September 28 of that year they would have to vote on the new Gaullist constitution, and they offer some recommendations that "represent the opinion of most of the members of the *Esprit* group—this review whose principal objective has been to elaborate the broad outlines of a new civilization but who are constantly called back to everyday politics by the French tragedy [i.e., the Algerian war]."[112] Four years later *Esprit* was at last free to close the cycle, to ponder what the new European and global civilization would and should look like. The theme of the July–August 1962 issue was *la planification*.

*

Returning once more in October 1962 to the subject that had obsessed him since 1954, Jean-Marie Domenach observed in "After the War" that memories were fading already. "The modern world possesses a formidable capacity for forgetfulness."[113]

Domenach's words are echoed a decade later, in March 1972, by I. F. Stone, who as the war in Vietnam was winding down, wrote in the *NYRB* of the "happy amnesia" that afflicted American leaders. A crucial role for the engaged intellectual, "if we are ever to disentangle ourselves from Indochina, is to force the painful record back into public consciousness. The facts are well-known, but constantly forgotten."[114]

One quality that all historians would ascribe to themselves, no matter what their ideological formation, is a congenital distaste for forgetfulness.[115] From this perspective we owe a debt of gratitude to Jean-Marie Domenach and his associates at *Esprit*, primarily Catholics but also Protestants, Jews, and unbelievers. And from the narrower perspective of our study of cycles of intellectual engagement, the *Esprit* group also commands our admiration. It was no small triumph, especially in a moment of national crisis, to remain faithful to Albert Camus's twin engagements.

Postscript: Why Catholic Engagement?

In the preceding discussion we have deliberately shied away from psychobiography, which in the study of intellectuals and the motivations for their engagement usually amounts to little more than a simpleminded "absolution of guilt for having a comfortable life" theory. We shall continue to do so in analyzing other intellectual engagements. In the case of Domenach and the important group of engaged liberal or left Catholic intellectuals he represents so well, there is a straightforward historical explanation that merits our attention. It is important to remember, especially because the memory has never left them,[116] that Domenach's generation lived through the Second World War and the humiliation and terror of the Nazi occupation and were old enough to have some recollection of the rise of fascism and the Spanish civil war. In his autobiography Domenach offers as persuasive an explanation for the consistent and faithful engagement of the *Esprit* group during the long years of the Algerian War as I have seen. It derives clearly from what I would call an "injured" historical consciousness. Domenach wrote that he always carried with him a photograph dating from 1937 or 1938, showing a group of Spanish archbishops standing on the steps of a cathedral, "forming with trembling arms the fascist salute to greet Franco's generals who were coming to celebrate their victory. Catholics of my generation have had to live with that shame. But in a certain manner, it whipped us into action, it incited us to combat."[117]

Tiersmondisme: Jean-Paul Sartre and *Les Temps modernes*

In a brilliant retrospective essay, looking back at the French Resistance to the Algerian War more than twenty years after the Évian accords, Pierre Vidal-Naquet discerned three major ideological and political temperaments within the larger group of engaged intellectuals: the Dreyfusards, the Bolsheviks, and the Tiersmondistes.[118] As a good historian, Vidal-Naquet is cautious, admitting that his categories are "ideal types" and that an individual intellectual might be placed in all three. Yet his ingenious classification does stand up under analysis.

Though he is careful not to locate himself in one of his categories, Vidal-Naquet belongs, I think, with the Dreyfusards, the group of intellectuals that we have linked especially with *Esprit* and whose engagement included a "French and even patriotic" dimension. For these intellectuals it was intolerable that the "country of the Rights of Man could tolerate, even order and organize torture and massacre."[119] Vidal-

Naquet dedicates his essay to two members of the *Esprit* group, Robert Bonnaud and Paul Thibaud, rather than to Jeanson or Sartre, for example. The phrase that Vidal-Naquet chose as an epigraph to his article, Une Fidélité têtue (a stubborn fidelity) is taken from a 1957 statement by Bonnaud, published in *Esprit* just after the latter's return from military service in Algeria, and it is in the purest Dreyfusard tradition:

> Having seen close at hand the terrible suffering of an oppressed people, having participated in the unjust war that is being carried out against this population, I am left with a kind of stubborn fidelity to our own [national] values, which we ourselves have violated, a fidelity that one will excuse me for not calling treason.[120]

The Bolsheviks (understanding Bolshevism in the context of the Algerian War as pure revolutionary ideology unsullied by Stalinism, desirous of returning to the "radical hopes of the October Revolution," later betrayed in Soviet praxis) would include the Jeanson network and associated groups. Some of these groups wanted to inject a Marxist ingredient into the Algerian revolution; other groups, like Jeanson's, thought that the FLN itself represented the genuine Bolshevik tradition.

Ali Haroun, one of the five (and later three, when two were arrested), principal FLN leaders in France, spoke admiringly of Jeanson; his was "a courageous engagement, the full lucidity of which has been shown by the course of history."[121] Jeanson's engagement (or, as I have suggested, *embrigadement*) was enormously helpful and greatly appreciated by the FLN leadership. For example, as late as 1961, 80 percent of the funds for the FLN cause came from Algerian workers in France, and all that money had to be conveyed secretly across the French border. Jeanson's network was the principal, though not the only, group involved. The transfer of funds out of France was so efficient and well organized that during the four years the network was in operation, only one of Jeanson's couriers absconded with his suitcase full of money.[122] The actions of these new Bolsheviks have been brilliantly recounted by Ali Haroun and Hamon and Rotman and so will not be examined in detail here. Even if we were to agree that the suitcase carriers were truly *engagé*, the form of counterlegal engagement that they chose was so dramatic and secretive—Hamon and Rotman's and Ali Haroun's works read at times like spy novels—that there was little room or need for the peculiar attributes we associate with intellectuals. The deeds of these clandestine group are, I believe, best understood and studied as political acts that could have been carried out by any courageous and determined individual, with or without training as an intellectual.

Turning now to Vidal-Naquet's third group, some of the Tiersmondistes were religious, but according to Vidal-Naquet the secular heart of the movement, especially in the last three years of the Algerian War, lay with Jean-Paul Sartre and *Les Temps modernes*.[123]

The link between Sartre and Tiersmondisme, at least at this stage in his career, has been widely recognized. He and Simone de Beauvoir traveled extensively in the Third World, including famous excursions in 1960 to Cuba and Brazil. Sartre greatly admired Frantz Fanon, and his 1961 preface to the *Wretched of the Earth*, a work that has sold over a million copies, is "furiously tiersmondiste," and perhaps is the most influential such text ever written by a European.[124]

By January 1956, when Sartre delivered a major speech, "Colonialism Is a System," at a public meeting in Paris organized by the Action Committee of Intellectuals Against the Pursuit of the War in North Africa, a principal ingredient in his Third World ideology was a belief in the inevitability and the justice of Algerian independence.[125] By 1957 Sartre was marching for peace along with François Mauriac, and he did not hesitate to sign the Manifesto of the 121 in September 1960. No study of French intellectual engagement during the

Algerian War can or should ignore Sartre's formidable, if somewhat tardy, contribution. The timing of Sartre's engagement, specifically its lateness, is interesting and quite puzzling, and its roots deserve a more detailed examination.

A Slow Awakening

Two months after the cease-fire in March 1962, Jean-Marie Domenach and the mathematician Laurent Schwartz of the Comité Maurice Audin took the initiative of collecting signatures on a petition for amnesty for draft resisters and other antiwar activists, including groups like the Jeanson network. The intellectuals who signed this petition were quite extraordinary in their diversity, ranging from the conservative Robert Aron to the former minister of aviation in the Popular Front government, Pierre Cot, François Mauriac, Catholic and Protestant leaders, Communist party members, cinematographers like Alain Resnais, scholar-activists like Pierre Vidal-Naquet, and Simone de Beauvoir and Jean-Paul Sartre. The petition, brief, a model of clarity and elegance, was published in *Esprit*, not in Jean-Paul Sartre's *Les Temps modernes*.[126]

Once again, this time at war's end, Sartre—unavoidable in any serious discussion of contemporary engagement, with a global reputation, incomparably better known than Domenach, famous for his radicalism, despised by some and honored by others for his unflinching atheism, and author of a play, *Le Diable et le bon Dieu*, that literally puts God on trial—was upstaged by *Esprit* and the Catholic Left. Sartre has often been accused of intellectual dishonesty, but in "Une Victoire" of March 1958, one of the finest occasional pieces in his long career and one of the most powerful denunciations of torture ever written, he gives credit where it was due. Until Henri Alleg's *La Question*, which had just been published and then banned by the authorities, "those who dared to bear witness were draftees, and especially priests."[127]

Les Temps modernes (henceforth *LTM*), which Sartre had

edited from its creation in 1945, was and is an internationally recognized periodical that by its very title, echoing the Charlie Chaplin film, is committed to dealing with contemporary issues. Sartre's name is so indissolubly associated with *LTM* that *Ulrich's International Periodicals* still lists him as director eight years after his death!

Going back to the very beginning of the Algerian struggle, it is truly perplexing that *LTM*, which any student of French intellectual life of the post-1945 period would expect to have been more *engagé*, more alert, and more radically anticolonialist than *Esprit* was, does not even refer, and then only briefly, to the Algerian War until May 1955. The position taken then was a moderate one, rather akin to that of Sartre's archrival Camus.[128] There is no thorough discussion of the conflict until October 1955, when it had been raging for almost a year, and ten months after Domenach's prophetic and already *engagé* article, "Is It War in North Africa?" was published. This paradoxical fact has been overlooked by Sartre scholars and by those who have studied the Catholic Left.[129] The section of Annie Cohen-Solal's definitive biography that covers the period from 1956 until Sartre's death in 1980 is entitled "Un homme qui s'éveille," and she offers a brilliant portrayal of a man whose "awakening" began at age fifty-one. But she does not compare Sartre with other intellectuals who rose sooner from their slumbers.

Nor, obviously, has anyone dealt with the question of why Sartre was such a late riser compared with other intellectuals such as Domenach. A tentative answer may lie in the global focus of *LTM*, which gradually adjusts to become "Tiersmondisme." To illustrate briefly—in the April 1955 issue there is an article by Marcel Willems entitled "A Balance Sheet of French Colonization: The Economy of Black Africa," and the September issue (the same month that *Esprit* published Sarrazin's predictions regarding an independent Algeria) includes a report entitled "Scenes of Repression in Venezuela," but there is nothing on the Algerian War in either issue.

LTM began to address seriously and quite dramatically the issues raised by the rebellion in Algeria in the lead editorial, "Refusal of Obedience," in the October 1955 issue, which skips the pedagogic stage completely. Although it has a strong moral tone, it is already close to the counterlegal position.[130] Until 1957 and the battle of Algiers, *LTM* continued to deal with the Algerian war only sporadically. Following this point a little further, to give a sense of the range of interests in *LTM*: The lead article in the January 1956 issue is entitled "A Future for the Antilles," and the issue includes a piece on American universities by Tito Gerassi, but nothing on Algeria. The February issue has a major article by Sartre entitled "Reformism and Its Fetishes," dealing with European communist parties, but again nothing on Algeria. There is also a piece on the Jewish question and Israel's right to existence and an article on Franco's Spain by Elena de la Souchère, probably the best-informed specialist on contemporary Spain living in France at the time.

Finally, the breadth of vision of *LTM* is well illustrated in the May 1956 issue, which does not examine the Algerian question despite the growing crisis there following Mollet's February 6 speech in Algiers and the March 16 vote of special powers.[131] The issue includes a poem honoring Emmett Till, an article on the Twentieth Party Congress in the Soviet Union, one on socialism in Yugoslavia, one on the activist reformer Danilo Dolci and his work in Sicily, an extensive debate on the state of Israel (with a communication by a moderate Arab spokesman), and a long discussion of the realities of Communist China.

Sartre's War

After 1959, and especially after the fall of 1960 and the uproar over the Manifesto of the 121, in a very real sense the Algerian War became *la guerre de Sartre*. This phrase, quoted by Annie Cohen-Solal, if taken out of context would appear bizarre if not demeaning, given what we know already about the Algerian War. It was meant in all seriousness, however; Sartre had completely missed the Spanish civil war and he had basically ignored the Popular Front. He was barely involved in the Resistance between 1941 and 1944, for complicated reasons, in part because he was not fully trusted by Resistance leaders, given the fact that he had been released from a German POW camp in 1941. "He would then have missed them all, every major political event of his epoch, except this one, the Algerian War. And this was, in a certain sense, the joining of a great cause with a great personality."[132]

Once the full significance of the Algerian War became apparent to Sartre, de Beauvoir, and the group of writers associated with *LTM*, they dealt with it extensively in their journal. Between the passing reference in May 1955 and Francis Jeanson's overview of the newly independent Algerian state, published in the November 1962 issue, there are 148 articles in *Les Temps modernes* dealing with the Algerian conflict, and from 1957 to 1961 the coverage is extremely thorough. As the guiding spirit behind *LTM* Sartre channeled much of his amazing energy and intellectual power into the struggle to end the war.

The intellectual ingredient in Sartre's engagement, his writings on the war, such as "You Are Super" (Vous êtes formidables) in 1956[133] and "We Are All Assassins," written early in 1958 just after the bombing of the Tunisian village of Sakhiet by the French air force,[134] are superb examples of polemical journalism, with a powerful philosophical and sometimes sociological overlay—Sartre was a very keen observer of trends in the media. These articles deal unsparingly with issues of collective guilt and thus the historical parallel with the Nazi years, torture, war crimes, and the danger of fascism.

Although the moral ingredient remained a strong element in their engagement for the duration of the war, by 1959 Sartre and *LTM* had explicitly endorsed counterlegal action, and Sartre was determined to involve himself in such action. As early as 1958 the pages of *LTM* were open to the most articulate and probably the most militant of the draft resisters, Maurice Maschino, publishing in the October issue an important excerpt from *Le Refus*, which openly advocates the refusal of induction, or desertion if one is already in the army, and exile.[135]

LTM published many other inflammatory texts, including in the spring of 1960 a letter from Jeanson to Sartre that offered an ardent defense of the former's clandestine activities on behalf of the FLN.[136] Jeanson makes a very militant statement, critical of the cowardice of the official Left, noting that when certain intellectuals persist in 1960 in telling "those who have chosen to engage themselves totally with us" that they were wrong in not limiting themselves to legal methods, "I do not know if one should conclude that they [our critics] are criminals or imbeciles."[137] In the summer of 1960 *LTM* published a report of the first clandestine congress of Jeune Résistance, a group of draft resisters separate from the Jeanson network, with the primary mission of helping deserters and those who refused induction leave the country and find jobs while in exile.[138]

LTM provided extensive coverage of the major political trials resulting from the war, especially that of the Jeanson network, which lasted from September 5 to October 1, 1960. It published many reports by the flamboyant and controversial Jacques Vergès, one of the chief attorneys in the Lawyers' Cooperative that defended both Algerian FLN prisoners and their European supporters. Vergès later became famous (or infamous) as the defense lawyer for Klaus Barbie and attempted to use arguments drawn from his experience during the Algerian War in Barbie's Lyons trial in 1987.

There were a number of disruptive incidents during the Jeanson trial, and Vergès was finally suspended and not allowed to present his closing arguments. Because their publication *in extenso* was also forbidden, Vergès turned to the pages of *LTM* to review the lessons of the trial. He admitted that the heart of the event was a propaganda battle, an effort to get the truth circulated. In a conclusion that could easily have been drafted by Sartre, Vergès wrote, "From the denunciation of torture to the proclamation of the right of insubordination the path is direct [*la route est droite*]."[139]

The Manifesto of the 121

Let us follow that path to Sartre's counterlegal engagements, which really began in the spring of 1959 when a secret meeting was arranged between him and Francis Jeanson. On that occasion Sartre put himself totally at his former associate's disposition.[140] Since 1956 Sartre and Jeanson had been involved in an ideological dispute over the attitude to take toward the Soviet repression in Hungary, and Jeanson was surprised and delighted at receiving Sartre's spontaneous and immediate support, which he greatly valued. From a public relations viewpoint this was a real coup: "Having Sartre on one's side meant gaining the allegiance of thousands, indeed hundreds of thousands, of followers."[141]

Given Sartre's enormous fame and extreme visibility—with his short stature and his crossed eyes he was easily recognizable—he never carried a valise, though he volunteered publicly to do so in the fall of 1960.

Without a doubt the most notorious of Sartre's engagements during the Algerian War was his signing of the Manifesto of the 121 in September 1960. This text has been mentioned on a number of occasions, including in Chapter 2 as an inspiration for the "Call to Resist Illegitimate Authority" in 1967. We also have noted the generally negative impact it had on Catholic antiwar intellectuals—only two well-known Catholics signed it, Robert Barrat and André Mandouze. The furor that the manifesto caused was probably aggravated by the fact that its precise wording was not immediately known, as its printing was banned by the government. The manifesto never appeared in *Le Monde*, and the pages of *LTM* that were to have included it were conspicuously blank. The complete document became briefly available in France only in 1961, when it was published in *Le droit à l'insoumission*, a collection of texts dealing with the controversy, edited by François Maspero. This volume was promptly seized by the government.[142]

In *Le Monde* during September and October 1960 there are fascinating brief references to what must have appeared to many readers as a mysterious document. When a famous intellectual figure such as André Schwarz-Bart or Françoise Sagan added his or her name to the list of signers, *Le Monde* took note. Journalists also reported on the sanctions taken by the government against some of the signatories, and *Le Monde* published a list of the 180 who had signed through September 30, 1960, including Clara and Florence Malraux, the ex-wife and daughter of de Gaulle's minister of culture.[143]

Le Monde printed in its entirety a countermanifesto of October 1960 that condemned the work of "the professors of treason," accused of being a "fifth column" that draws its inspiration from "foreign propaganda." This manifesto was signed by nearly three hundred intellectual supporters of Algérie française, including seven members of the French Academy.[144] But at the time readers could only speculate as to the exact nature of the "treason" supposedly perpetrated by these "professors."

The subject of all this agitation was a three-page document, divided into five sections. The first four set the background with a brief history of the Algerian War.

> Neither a war of conquest nor a war of "national defense," nor a civil war, the Algerian War has gradually become an action proper to the army and to a caste that refuses to cede when confronted with an uprising that even the civil power in France, coming to recognize the general collapse of colonial empires, seems ready to comprehend.

The manifesto continues with a reminder that fifteen years after the defeat of the Third Reich, "French militarism, as a result of the demands of this war, has managed to restore torture and to make it once again practically an institution in Europe."

The manifesto takes note of the questioning that is becoming more and more prevalent among young French people, and the growing revolt against the army, including more and more acts of insubordination, desertion, and aid to Algerian combatants. A new rather spontaneous "resistance" has been born, which transcends traditional political groupings and is not understood by the official press. Then the formal and binding declarations follow:

> The undersigned, considering that each individual must make up his own mind regarding acts that it is henceforth impossible to present as isolated incidents of individual adventuring; considering that we ourselves, in our particular situations and according to our means, have the duty to intervene—not to give formal advice to men who have to arrive at personal decisions when confronted with such severe problems but to demand that those who are judging these individuals [draft resisters] not allow themselves to find anything equivocal in their statements and values, declare:
>
> • We respect and deem justified the refusal to take up arms against the Algerian people.
>
> • We respect and deem justified the conduct of Frenchmen who esteem it their duty to supply aid and protection to Algerians who are oppressed in the name of the French people.
>
> • The cause of the Algerian people, who are contributing in a decisive manner to destroying the colonial system, is the cause of all free men.

Sartre did not return from his trip to Brazil until early November 1960, but his presence while abroad had been felt in

France. Besides signing the Manifesto of the 121 an inflammatory open letter dated September 16, 1960, was prepared over his signature for the trial of the Jeanson network and read in open court. It was published in *Le Monde* on September 22, 1960. Sartre's provocative letter was a deliberate challenge to the authorities, practically begging them to arrest him. It stated emphatically: "If Jeanson had asked me to carry valises or to shelter Algerian militants, and if I could have done it without risk for them, I would have done so without hesitation." The suitcase carriers "represent the future of France, and the ephemeral power which is preparing to judge them no longer represents anything."[145]

If Sartre were not France's most famous intellectual and the tremendous media figure that he had become, his counterlegal engagement of 1960 would surely have become *embrigadement*. But Sartre never was arrested and was hardly bothered by the police. De Gaulle was much too shrewd to create a martyr and treated the Sartrians and the intellectuals in general, as long as they were not state functionaries, with a bemused and tolerant caution. "One does not imprison Voltaire."[146] There is, however, no question of Sartre's sincerity or his courage; his apartment and the offices of *Les Temps modernes* were bombed with plastic explosives, and militant supporters of Algérie française marched through the streets of Paris in October 1960 calling for his assassination.[147]

On the day after the cease-fire, March 19, 1962, Sartre penned "The Sleepwalkers," his bitter reflections on the war just ending. He certainly takes no credit for his engagement, and the general tone is one of regret. Echoing a plea that had been made two years earlier by Jean-Marie Domenach, Sartre argued that if only the Left had been able to surmount its divisions, many lives could have been saved. The real defeat of the Left, Sartre writes, is the million Algerians we have allowed to be killed. We gave the whole affair over to de Gaulle in 1958, and so the Algerians have conquered their liberty while the French lost theirs. The French are "going as sleepwalkers toward their destiny." Now they have arrived, eyes closed, at the crossroads. The tasks of the Left are to force the government to prosecute the OAS, to maintain the loyalty of the army, and to unite to guarantee the implementation of the Évian accords. "Under these conditions the cease-fire for us could be the beginning of a beginning."[148]

Sartre did not participate in the ceremonies of July 1, 1962, celebrating Algeria's formal independence. For him the cycle of engagement was complete. He had returned to more purely intellectual tasks and was deeply involved in completing his autobiography, *The Words*, which is, in the view of many, his masterpiece.

The End of the Beginning

We have already noted the extreme rapidity with which France changed once the Algerian War finally was settled. Whether the Algerian episode produced in France the last of the internal wars, *les guerres franco-françaises*, that have periodically split the French nation like "geological faults," with periods of calm between fractures, remains to be seen.[149] Certainly nothing since the resolution of the Algerian struggle has divided the French so bitterly as did their last colonial war. As John Talbott pointed out, other social conflicts did emerge after 1962, some of which had been obscured or postponed by the war itself. But none of these disputes "involved issues over which the French could easily contemplate killing each other."[150] Talbott's observation is pertinent and convincing, especially when one realizes that no students were killed during the chaotic upheavals in Paris in May 1968.

In the immediate wake of the Algerian War, Michel Crouzet published an influential retrospective essay entitled "The Battle of the French Intellectuals." Crouzet had been a member of the French Communist party but left largely because of its hesitant and dilatory attitude toward the war in Algeria and decolonization, and he was himself active in the antiwar moment as an independent leftist. Crouzet gives a lot of credit to professors who were "particularly firm in their scientific analyses of the rise of nationalism in the colonized countries and in their affirmation of the fatality of independence."[151] He argues forcefully that for the intelligentsia, the years from 1954 to 1962 were a period of mobilization and, in a frequently cited phrase, a "battle of the written word" (*bataille de l'écrit*). Crouzet believed that the antiwar intellectuals had won this battle hands down. "The intellectual 'class' was massively in a state of protest against the Algerian War."[152]

Crouzet's view is largely supported by historians and sociologists looking back a quarter-century later. Bernard Droz observed that although the cause of Algérie française received the support of some talented intellectuals, rarely has the French intellectual class "been so united (*soudé*, literally "soldered" together) than in its denunciation of the Algerian cancer."[153]

Once that cancer had been excised from the French body politic, and the battle of the word won, the solder melted, the energies of the French intellectual class dispersed, and there was a nearly instantaneous, widely noted, and sometimes lamented *dégagement*.

Will ever be another *bataille de l'écrit*? Or have socioeconomic changes, especially the power of television and the growing functional illiteracy found even in France, render impossible a resumption of the battle and thus a new manifestation of the cycle of engagement? This double-edged question can best be reflected on after we have examined the American variant of the cycle of engagement during the Vietnam years.

Jean-Paul Sartre thought that the cease-fire accords of March 1962 might signal the "beginning of a beginning." Many specialists on contemporary France and the history of its intellectuals believe that the independent New Left of the later 1960s was born out of habits learned and hopes both raised and shattered by antiwar engagement during the Algerian conflict. If this view is correct, from the perspective of the political and social history of Sartre's native land, the end of the beginning would be the "events" (*les événements*) of the spring of 1968. This largely leaderless student "revolution," centered in Paris, for an euphoric moment seemed capable of sweeping aside the Fifth Republic and its president, Charles de Gaulle, in a wave of happy anarchy, before dissipating as swiftly and mysteriously as it had appeared.

From the perspective of intellectual history and the study of cycles of engagement, the end of the beginning may perhaps be found in the December 17, 1964, issue of the *New York Review of Books*. The reason for this precise localization of a phenomenon that would seem difficult to pinpoint is that Jean-Paul Sartre did in fact reconsider the Algerian War in 1964 after the regime in Algiers had stabilized, and in retrospect he was less pessimistic about his engagement.

The *NYRB* published an English translation of Sartre's statement to the Swedish press at the time he refused the Nobel Prize, seven years after Albert Camus accepted that award. Sartre stated his general policy of declining official honors, noting that even though his sympathies were with socialism, he would refuse the Lenin Prize if it were ever offered to him. However, there was an exception:

> During the war in Algeria, when we had signed the "declaration of the 121," I should have gratefully accepted the prize, because it would have honored not only me but the freedom for which we were fighting. But matters did not turn out that way, and it is only after the battle is over that the prize has been awarded me.[154]

Sartre's communication appeared in the very same issue of the *NYRB* as did the D. A. N. Jones article cited in Chapter 1 as the first example of Algeria–Vietnam parallels, and just as the *NYRB*'s long obsession with Vietnam was beginning.

4

Vietnam: The Acid Test of an Intellectual Generation

America's Vietnam War has cast its ugly shadow over much of what has come before in this book. Comparison between the Algerian and Vietnam wars was the specific focus of Chapter 1. Chapter 2 offered a theoretical interpretation of the cycles of intellectual engagement that both wars generated, and it illustrated this interpretation with examples drawn from the Vietnam era. Chapter 3 began and ended with a look ahead to the American intellectual response to our Vietnam War. We shall now examine that response in greater detail.

Bertrand Russell once claimed that "Vietnam is an acid test for this generation of Western intellectuals."[1] If opposition to the Vietnam War is viewed as the positive side of the test, Western intellectuals—Americans and their European peers—passed with flying colors. Had the intellectuals decided it, there would never have been a significant American military involvement in Vietnam. From its inception as a full-scale conflict after the Gulf of Tonkin Resolution in August 1964, the war was opposed by a majority of the intelligentsia on both sides of the Atlantic.

As was the case with the French intellectuals and their engagement during the Algerian War, our principal concern here will be intellectuals with relatively wide public recognition, who were also American citizens. This was the group that inevitably felt most directly responsible and implicated, because it was their nation that had intervened and then escalated the conflict in Vietnam. Among the American intelligentsia the degree of opposition increased as the war escalated, to the point that it would be safe to say that by 1967 it was overwhelming, at least among intellectuals affiliated with elite educational institutions primarily located in the Northeast and California, and the country's most famous independent novelists, essayists, artists, and poets. There is, we shall see, convincing evidence of the truth of this assertion, but to carry it further and make estimates regarding the percentages of the broader intellectual class, including students, opposed to and supporting America's war in Vietnam is problematic at best.

Counting

Of course the right lobe of the brain was active in the American debate over Vietnam between 1964 and 1975, as it had been in France during the Algerian War. There is important research to be done on the "prowar" minorities in the intelligentsia in both countries. (It is perhaps more accurate to say "progovernment"—supporting the wars as necessary evils leading to greater goods.) Illuminating comparisons will probably emerge. During the Vietnam War, especially between 1964 and 1968, manifestos supporting our involvement were circulated and signed by such important American intellectuals as Sidney Hook, Max Lerner, John Dos Passos, William Buckley, Lucien Pye, and Ithiel de Sola Pool. It is a fact that these petitions were fewer in number and in number of signatories than were those drafted by the antiwar intelligentsia. It is also a fact that the progovernment intellectuals such as those just cited did not and do not command the respect *qua* intellectuals of their antiwar *confrères*. Whether this perception is based on true and fair judgments (which, after all, are largely made by other intellectuals) is not at issue here. Its existence would not be questioned by intellectuals on all sides of the political spectrum.

The question of numbers more broadly construed, including several million college and university students and their professors, is a vexed one, especially when we link it in with more subjective issues of influence and importance. It will probably always remain so. What we conclude hinges on whom we wish to assign a place in the intellectual class and how one measures influence and importance. Our democratic instincts might make us wish to assign equal weight to a prowar statement by a poet in residence at a small western college (if such a text exists), as to Robert Lowell's antiwar activity, which was heavily covered in the *New York Times*. Common sense does suggest that the latter had more impact than the former.

As with almost every other topic pertaining to America's involvement in Vietnam, there is an extensive specialized literature attempting to measure opposition to the Vietnam War among the larger intellectual class. Some of it is useful and interesting, though the results are often inconclusive. The manner in which university faculties were sampled and the number of respondents leave much to be desired.[2] But this research does show convincingly that there was a difference in attitudes and participation between the elite and the less famous Catholic and southern and midwestern colleges and universities. "Anti-war sentiment and anti-war activity was concentrated at the more prestigious (and visible) institutions."[3] Fewer agriculture, business, and education professors opposed the war than did those in the humanities and social and natural sciences.

The high level of opposition in America's most famous universities is indisputable. On October 7, 1969, the Harvard faculty voted 255 to 81 against the American military involvement in Vietnam, and 391 to 16 in support of the planned October 15 Moratorium Day of debate and discussion of issues raised by the war. This was the first and last such public stance taken by Harvard's faculty in the university's 354-year history.[4] It is also true that that even government officials recognized that most of the talent lay with the opposition. After a particularly strong and influential antiwar article by David Halberstam appeared in the December 1967 issue of *Harper's*, one of Walt Rostow's aides lamented, "I wish we had people on our side who could write like he does."[5]

The central focus of this book remains the intellectual elite, and one of the most useful sources in gauging this group's opinion on the Vietnam War is *Authors Take Sides on Vietnam*, edited by Cecil Woolf and John Bagguley and published in 1967. To gather material for their book, Woolf and Bagguley simply wrote to more than three hundred well-known authors from several countries, requesting brief answers to the following questions: "Are you for, or against, the intervention of the United States in Vietnam?" "How, in your opinion, should the conflict in Vietnam be resolved?" They received more than 250 replies, an amazingly high percentage, as anyone who has ever administered a questionnaire knows, "a response from writers which has had no precedent in the events of the past twenty-five years."[6]

The American contingent numbered sixty-four. Of that group fifty-five were against American intervention in Vietnam, five were in favor, and four could not make up their minds. The list of those opposed includes many of the leading intellectuals of the day, and their comments range from the satirical to the despairing; many are informed by a sense of bitterness, helplessness, and moral outrage. Several of the respondents, including Gore Vidal and Kay Boyle, spoke of their shame at being American.[7] Nelson Algren, among others, brought up the Nuremberg precedents, advocating bringing President Johnson to "public trial for crimes against humanity."[8] Four drew the analogy with the French and Algeria, among them S. N. Behrman, who in arguing for American withdrawal from Vietnam claimed that neither de Gaulle nor France lost face by the withdrawal of 1962.[9] Although mathematical proof is obviously impossible, I am convinced that the percentages derived from these numbers—roughly 90 percent opposed, 5 percent in favor, and 5 percent undecided, approximate the attitude held toward the Vietnam War by the elite intelligentsia at large. The figure of 90 percent opposed to American intervention in Vietnam is identical with that found in 1970 among the leading American intellectuals surveyed by Charles Kadushin in his well-known study *The American Intellectual Elite*.[10] We can be certain about the poets. As W. H. Auden, who had become an American citizen and knew his craft and its practitioners as well as anyone, wrote in 1971, "Thank God, there is not a living American poet of major importance who openly supported our intervention in Vietnam."[11]

Petitions

In the words of Pascal Ory and Jean-François Sirinelli, "The petition is in truth the *degré zéro* of intellectual engagement."[12] We shall define *petition* broadly here, as a formal written document signed by a number of individuals, which may be a request from higher authorities, in the original Latin sense of *petitio*, or which may be a statement of principles or goals, perhaps a call to action often directed against political authorities who may be deemed illegitimate. Analysis of petitions, their changing language and themes, is an especially useful way to gain insight into how the cycle of engagement played itself out in the United States during the Vietnam War. The French case, which was part of recent historical memory, offered precedents and lessons, and many of the engaged American intellectuals were aware of the "war of petitions" that occurred in France during the Algerian conflict, and they even directly imitated the most famous of those manifestos, that of the "121."[13]

The number of antiwar petitions that were circulated during the Vietnam years was enormous—a veritable "flood" in the *New York Times* alone, in the words of the sociologist Everett Carll Ladd.[14] I cannot even begin to imagine approximate totals or how to arrive at them, as there were so many different sponsors, both regional and national, preparing texts that were directed to a wide variety of authorities and potential supporters. By 1970 organizers for the "National Petition Campaign" to protest the invasion of Cambodia felt the need to provide a lengthy and somewhat apologetic and defensive explanation as to why yet another petition was appropriate.[15] Viewed in chronological sequence these antiwar petitions reflect almost perfectly the cycle of engagement during the Vietnam years. Selected for analysis are ten representative petitions from the Vietnam era, with the seventh, the "Call to Resist Illegitimate Authority," being by far the most famous.

In its February 16, 1965, issue, the *New York Times* printed a petition in the form of "An Open Letter to President Johnson on Vietnam." This text was signed by 424 academics, representing twenty-six institutions of higher learning, all, except the University of Rochester, located in New England. Sixty-six academics were from the Massachusetts Institute of Technology, and 62 from Harvard, together comprising 30 percent of the total. One might have expected the large number of Harvard signers, and from President Johnson's own perspective he had some justification for his antieastern and especially anti-Harvard sentiments. The even larger group from MIT, then and now one of America's premier technological institutions and normally very conservative, is more surprising. Noam Chomsky was already on the linguistics faculty there, but his name does not appear on this petition; he had not yet become fully *engagé*, galvanizing his colleagues with the passion and brilliance that later were to make him America's best-known and probably most influential antiwar academic. In 1969 he regretted the relative lateness of his engagement: "No one who involved himself in antiwar activities as late as 1965, as I did, has any reason for pride or satisfaction. This opposition was ten or fifteen years too late."[16]

The reason for the large participation from MIT has to do with dashed hopes and betrayed expectations. Most of the signers had a few months earlier been members of "Scientists and Engineers for Johnson and Humphrey," which had actively campaigned in the 1964 presidential elections.[17] Their strong support for Johnson had been motivated largely by their fear of Barry Goldwater, the Republican candidate, who with his slogan "In your heart you know he's right," seemed irrational enough to order the use of nuclear weapons in Vietnam if elected.

The tone of this open letter to President Johnson is moderate and accommodating, very much in keeping with the first level of engagement, the *pedagogic*. The letter asserts that America has lost the initiative in Vietnam and that we are substituting military goals for political ones. It is framed as a series of polite queries to the president, including: "What are our goals in Vietnam? Are they just? Can they be accomplished?" The letter expresses worry about the possibly prohibitive cost of achieving these goals in terms of lives and money. There is also uneasiness about our ally, the unstable, unpopular, and undemocratic regime in Saigon.

This early letter even acknowledges that our initial motives in going into Vietnam after the French had been forced out in 1954 were fair and decent—to develop democracy there and to aid in the formation of a stable representative government. But "historical, political, social, religious and sectional factors have prevented this development. The original assumptions are no longer valid." The letter is so cautious and polite that it concedes that President Johnson might have some secret data he could not reveal, but "could such information completely refute the picture of events and the political insights provided to us by serious newspapermen who have been in the area for years?" It concludes with a call for a negotiated peace while hinting at the possible emergence of the moral dimension of protest: "If we are not to widen the war beyond all conscience, as reasonable men we must negotiate while there is still time."

By May 1965, the numbers of signatories were escalating along with the war. The Greater Boston Faculty Committee on Vietnam alone gathered 676 signatures from professors at thirty colleges and universities (200 from Harvard and 136 from MIT, including Noam Chomsky). They signed an open letter to Secretary of State Dean Rusk, published in the May 9, 1965, *New York Times*. The letter responds to Rusk's April 23 attack on academic critics of the Vietnam War, which the secretary of state had made in a speech to the Society of International Law. According to this rebuttal, Rusk's polemic had been unjustified and poorly argued. The administration's "stubborn disregard of plain facts" is emphasized, and the pedagogic tone of the letter is evident. The text may be read as a kind of lecture or lesson to those in power; our leaders, like our students, need to be enlightened and have their mistakes corrected.

Though America's preeminent position in the world is still assumed, the second level of engagement is explicitly prefigured in the letter to Rusk. The situation in Vietnam "raises serious moral questions, not merely diplomatic and tactical ones. As a nation we hold immense power. To permit it to be used in reckless and barbarous ways is to imperil the entire basis of American leadership."

The first antiwar petition published in the *New York Review of Books* was directed to its readership rather than to President Johnson or Secretary Rusk. (This petition also appeared in the *New Republic*; no single periodical held pride of place in circulating such documents to the public.) It was a call to attend an International Conference on Alternative Perspectives on Vietnam, which was held at the University of Michigan in September 1965. The document is especially interesting because as part of their appeal for wider support the initial sponsors of the conference review the success of the teach-in movement thus far and argue that it has been "extremely effective in raising fundamental issues and in analyzing the weaknesses and dangers of current policy." However, it has not yet identified and elaborated alternative policies, which is obviously also part of good pedagogy. Once these alternative solutions are found, our leaders must be told rather than advised what course should be followed to end the war.

Locating these new perspectives is a major challenge for the global intellectual community and requires the combined participation of "humanists and religious thinkers, of social theorists and social philosophers, of students of Southeast Asia and of the developing world." American intellectuals must, the sponsors argue, take the primary responsibility for meeting this challenge, because our government through its military involvement in Vietnam has created a "moral crisis."[18]

The second level of engagement is clearly present in a new communication to President Johnson that appeared in the October 31, 1965, issue of the *New York Times*. The polite tone of the previous February was abandoned, and 653 professors bluntly told the president to "stop the bloodshed in Vietnam." The increased suffering of the Vietnamese caused by our military presence is denounced, and the American people are themselves "undergoing a brutalizing and degrading experience. Americans are coming to view without concern the inhuman suffering being inflicted in their name against a defenseless population." The petitioners insist that Johnson order an immediate halt to the destruction of villages and the burning of food supplies, that is, end the policy of "pacification" and its resultant forced urbanization. In addition, all bombing and all offensive military operations must cease in order to permit the Vietnamese to work out their own peace settlement, as promised by the 1954 Geneva accords. The language is strong, but the emphasis is still on legality. The signers will "continue to exercise our lawful right to protest. We shall oppose your present policy in Vietnam openly and vigorously. Only thus can we do our part in trying to bring about peace and to protect the moral integrity of our nation."

By the third year of the war, 1966, the professors had given up on President Johnson, as had other segments of the intellectual antiwar movement. Our thirty-sixth president had become a subject of antipathy, mockery, savage satire, and devastating caricature.

As they turned away from their president, the professors looked directly to the American people and the Congress of the United States. In an "Open Letter on Vietnam" published in the *New York Times* on February 13, 1966, 1,263 academics representing seventy-one colleges and universities spoke out "in protest and in warning." They expressed deep distress that the New Year's truce of the previous month had not led to peace but, instead, to further escalation of the war. The moral tone is very strong: "U.S. bombs, napalm, and chemical warfare are making a desert of the country. The only peace that can be achieved in this way is the peace of the grave. We recoil in horror." The professors call for the immediate cessation of all bombing raids on North and South Vietnam and for complete American troop withdrawal, regardless of the type of government the South Vietnamese might choose following a negotiated peace settlement and genuinely free elections.[19]

"Individuals Against the Crime of Silence," which first circulated in June 1967 and was widely distributed during both the Johnson and Nixon administrations, was a different type of petition, addressed to "our fellow citizens of the United States, to the peoples of the world, and to future generations." This succinct text, only eleven sentences, was designed to gather large numbers of signatures from across the spectrum of antiwar opinion. Though its emphasis on moral conscience clearly places it at the second level of engagement and it is not counterlegal itself, it encompasses or at least admits the validity of all three levels of engagement. Its primary purpose, beyond expressing yet again anger and distress at "the conduct of our country in Vietnam," is to maintain a permanent historical record, and in this sense it may reflect the wounded historical consciousness that I have contended impelled many French intellectuals into engagement during the Algerian War.

According to its sponsors this important petition attracted 100,000 signers during the Johnson administration alone.[20] Its four key points are cited in their original wording:

> We, the signers of this declaration, believe this war to be immoral. We believe it to be illegal. We must oppose it.

> At Nuremberg, after World War II, we tried, convicted and executed men for the crime of OBEYING their government, when that government demanded of them crimes against humanity. Millions more, who were not tried, were still guilty of THE CRIME OF SILENCE.

> We have a commitment to the laws and principles we carefully forged in the AMERICAN CONSTITUTION, at the NUREMBERG TRIALS, and in the UNITED NATIONS CHARTER. And our own deep democratic traditions and our dedication to the ideal of human decency among men demand that we speak out.

> We therefore wish to declare our names to the office of the Secretary General of the United Nations, both as permanent witness to our opposition to the war in Vietnam and as a demonstration that the conscience of America is not dead.

A Call to Resist Illegitimate Authority

Planning for the single most influential petition of the Vietnam era began in the spring of 1967, directly inspired by the

Manifesto of the 121.[21] Once the number of signatures was deemed sufficient, the document was published in the October 7, 1967, issue of the *New Republic* and the October 12, 1967, issue of the *NYRB*, with a partial list of the signers attached. Among the 130 names selected were those of Philip Berrigan, Noam Chomsky, the Reverend William Sloane Coffin, Allen Ginsberg, Paul Goodman, Denise Levertov, Dwight Macdonald, Herbert Marcuse, Linus Pauling, Edgar Snow, Susan Sontag, and Dr. Benjamin Spock.

The "Call," addressed to "the young men of America, to the whole of the American people, and to all men of good will everywhere,"[22] is clearly though carefully *counterlegal*. The relatively brief text (nine numbered paragraphs) begins at the second level of engagement, noting that the signers share the "moral outrage" of Americans of draft age who are refusing to contribute to the Vietnam War in any way.

Historical arguments are then presented to demonstrate that American military involvement in the undeclared Vietnam War is unconstitutional and illegal under our own jurisprudence and also violates the Charter of the United Nations. The Nuremberg precedents are explicitly cited. According to the "Call," the types of actions committed by American troops in Vietnam—such as the destruction of villages, internment of civilian populations in concentration camps, summary executions of civilians—are those that the United States and its allies in World War II "declared to be crimes against humanity for which individuals were to be held personally responsible even when acting under the orders of their governments and for which Germans were sentenced at Nuremberg to long prison terms and death."

The signers believe that "every free man has a legal right and a moral duty to exert every effort to end this war, to avoid collusion with it, and to encourage others to do the same." However, they do not explicitly advocate draft refusal, recognizing the agonizing choices faced by young men confronted with military service in Vietnam. "Each must choose the course of resistance dictated by his conscience and circumstances." In language reminiscent of that of their French predecessors in 1960 ("we respect and deem justified the refusal to take up arms against the Algerian people"), the signers assert that each of the several possible forms of resistance to military service in Vietnam is "courageous and justified." Their language is somewhat more explicit than that of the "121" in that they go on to state that many of their number "believe that open resistance to the war and the draft is the course of action most likely to strengthen the moral resolve with which all of us can oppose this war and most likely to bring an end to the war."

The ninth and final point is an open call to "all men of good will," and especially to universities and religious organizations, to join with the signers "in this confrontation with immoral authority.... Now is the time to resist."

Many other petitions followed, though none had the impact of this dramatic and widely circulated endorsement of counterlegal activities.[23] "A New Call to Resist Illegitimate Authority," published in the November 20, 1969, issue of the *NYRB* spun off directly from the first. By the fall of 1969 under President Richard Nixon the "Vietnamization" of the war was beginning, but a negotiated settlement was only dimly on the horizon if in view at all. The first "Call" had concentrated on the war and the draft. The signers now broadened the scope of their concerns, stating that opposition to the Vietnam War must lead to "opposing the institutions that support and maintain it," such as imperialism, militarism, economic exploitation, and racism.

They are also openly counterlegal in giving their full support to young men who refuse to register for the draft or submit to induction, to soldiers who go AWOL, and to those who impede the operation of draft boards.

In a concluding note, the "New Call" warns potential signers that it is possible that the very action of signing this petition could be found illegal by the government and hence may place them in jeopardy of arrest and indictment.

*

To give a sense of the variety of petitions that built on the momentum generated by the "Call to Resist Illegitimate Authority," I shall discuss two of the most creative that followed it. Rather than blanket condemnations or broad philosophical statements, such petitions often focused on single issues related in some fashion to opposition to the Vietnam War, appealing for aid in electing antiwar candidates in local districts, for example.

Much imagination is displayed in "if a thousand men were not to pay their tax-bills this year" of February 1968.[24] The quotation was taken from Henry David Thoreau's *Civil Disobedience* of 1849, commenting on American involvement in the Mexican war. The appeal is simple and shows graphically the movement from the moral to the counterlegal stage of engagement.

- We, the undersigned writers and editors, believing that American involvement in Vietnam is *morally wrong*, pledge:
  1. None of us voluntarily will pay the proposed 10% income tax surcharge or any war-designated tax increase.
  2. Many of us will not pay that 23% of our current income tax which is being used to finance the war in Vietnam.

This pledge is followed by a careful warning to those who might wish to emulate the 444 intellectuals who had already signed it, that willful refusal to pay federal income taxes is a violation of the Internal Revenue Code and is punishable by fine and/or imprisonment.

Among the signers were many of the intellectuals discussed earlier in this book, and such diverse writers and editors as James Baldwin, Stanley Elkin, Richard Ellman, Leslie Fiedler, Henry Miller, Scott Nearing, Terry Southern, Gloria Steinem, and especially Thomas Pynchon. Pynchon, one of America's finest and most respected novelists, is well known for his extreme reclusiveness, and is probably the most private of American writers save for J. D. Salinger. It is indeed significant that over this particular issue and type of Vietnam protest, Pynchon was willing to break his silence and make this public commitment.

A "Petition for Redress of Grievances" was presented to Congress in May 1972. An enterprising antiwar activist had remembered that the United States Constitution (actually the First Amendment) guarantees the right of the people to "petition the Government for a redress of grievances." A petition was drafted, which first reminded the legislative branch of our government that two years after the repeal of the Gulf of Tonkin Resolution by Congress, the executive branch persisted in its defiance of the legislative branch by continuing the war. The 176 signers then stated:

> We petition the Congress to exercise its Constitutional authority of control over the armed forces by voting an immediate cessation of all air, ground, and naval operations in Indochina, and by ending all military and economic appropriations for a war the Congress did not vote for and the American people do not want.

This group of petitioners obviously believed that their act was strictly legal and constitutionally authorized. The government took a different view, however, and after the petition was presented to the Speaker of the House of Representatives, ninety-four of the signers were arrested for "unlawful assembly" and jailed.[25]

Maintaining the Historical Record

After a thoughtful and balanced review of the French intellectuals' campaign against torture during the Algerian War, John Talbott observes tellingly that "assessing the impact of political protest is one of the historian's harder tasks." Talbott was not able to discover any evidence that the revelations made in journals like *Esprit* and *Les Temps modernes* and in books like *La Question* affected broad segments of French public opinion toward the war and Algérie française. Indeed, it is possible that in some cases such revelations at least momentarily strengthened the resolve of those favoring vigorous pursuit of the war and maintenance of French Algeria. Speaking to a meeting of French veterans' organizations in July 1957, Resident Minister Robert Lacoste, whose ardent, forceful, and, for a brief period, highly effective support for Algérie française has been previously noted on several occasions, stated: "The exhibitionists of the heart and the intellect who have mounted the campaign against torture are responsible for the resurgence of terrorism, which has caused in Algiers in recent days 20 dead and 150 wounded. I present them to you for your scorn" (*Je les voue à votre mépris*).[26]

Talbott continues his discussion of the impact of intellectual protest during the Algerian War by drawing the parallel with Vietnam: "Protest did not help bring down a government, as protest against the war in Vietnam helped discourage Lyndon Johnson [in April 1968] from seeking reelection."[27] Talbott's language here is carefully qualified. Doubtless internal dissent more broadly conceived, and within that spectrum of activity the engagement of America's antiwar intellectual elite, affected American policy toward Vietnam and played some role in convincing Lyndon Johnson not to seek a second full term as president. Exactly how this process operated is not susceptible to any sort of definitive analysis. As Melvin Small, author of the standard work on the subject, *Johnson, Nixon, and the Doves*, puts it, "The question of how much attention was paid to opinion and how seriously to take presidential hostility to critics can only be answered impressionistically."[28] Small's book is impeccably argued and carefully researched. He is able to demonstrate convincingly that protest had an indirect impact, for example, playing a role in President Johnson's decision to speak at Johns Hopkins University in April 1965 in defense of his Vietnam policy and in the announcement of a bombing pause from May 12 to May 17, 1965. In another illustration of a kind of backhanded influence, Small argues that the resources and energies expended by government intelligence agencies in monitoring dissenters, which obviously could have been allocated elsewhere, "suggest again the growing importance of the antiwar movement."[29]

Focusing more specifically on the intelligentsia, neither Small nor I would go as far as Charles Kadushin did, who stated that "the elite American intellectuals no doubt affected the progress of the war in Southeast Asia."[30] In the matter of Johnson's still not fully understood decision to retire from political life and simultaneously de-escalate the war, my own view is that the crucial factors were the Tet offensive and the counsel of that proud man's most trusted advisers. (To be sure, the surviving members of the latter group whom Small interviewed admitted that they had been affected by internal dissent, so there would again be evidence of an indirect influence.)

President Johnson's intense antipathy toward the war protest movement is widely documented and cited in all the important biographies and in studies of the Vietnam War. It is at least arguable that on certain key occasions Johnson deliberately ignored protest and took steps, perhaps in part out of spite, to prove that antiwar activities would have the opposite effect from what was intended, though an effect nonetheless. An example is his escalating the war by authorizing bombing inside the city limits of North Vietnam's port of Haiphong for the first time on April 20, 1967, destroying a power plant; and on April 25 the closest air raid to Hanoi was approved, bombing a rail yard only 2¼ miles from the center of the city.[31] These decisions were taken immediately after the huge peace march in New York City on April 15, 1967, the largest single such demonstration during the Vietnam War.

Returning to the comparison with Algeria, John Talbott goes on to cite a number of French antiwar activists who became extremely discouraged, believing that their efforts had been for naught, as the Algerian War dragged on through the Fourth and into the Fifth Republic. Talbott concludes that these engaged intellectuals were too pessimistic regarding their own achievements. "A substantial body of French opinion refused to countenance actions carried out in pursuit of the French government's Algerian policy. Protest against these actions stands as part of the historical record."[32] So, too, does the protest against American actions in Vietnam, represented and reflected with special clarity in the petition campaign, stand as part of the historical record, whether or not Presidents Johnson and Nixon took note of it.

"The *New York Review* of Vietnam": ## "The Bible of Vietnamese Dissent"

Between 1964 and 1975 some degree of intellectual opposition to the Vietnam War was articulated in a wide range of American periodicals, including *The Nation, Commonweal, New Yorker, Dissent, Harper's, Atlantic Monthly, Partisan Review, New Republic,* and even *Ramparts*, though the latter tended to feature exposé journalism with fewer intellectual pretensions.[33] Close comparative study will doubtless show variations in the level of antiwar sentiment to be found in these and other periodicals.[34] In none of them, however, even *Ramparts*, was the concentration on the Vietnam War as extensive, as intense, and as sustained as in the *New York Review of Books*. There are many reasons for this, including ideological underpinnings, intended audience, the niche in the intellectual field the periodical was designed to fill, aesthetic concerns, and editorial policy regarding week-by-week coverage of general news and the publication of fiction.

For the study of intellectual engagement during the Vietnam War, the *New York Review of Books* is far and away the most important single source. It was after all the periodical that became known in the late 1960s as "the New York Review of Vietnam" and for good reason, given the depth of its antiwar commitment.[35] Tom Wicker of the *New York Times* sharply if somewhat infelicitously characterized the *NYRB* in 1967 as "the Bible of Vietnamese Dissent."[36] Although the *NYRB* did publish a few communications from Vietnamese dissenters, such as the Buddhist poet Trich Nhat Hanh in 1966,[37] Wicker would have been more accurate if less trenchant had he written "the Bible of that majority of the American intellectual class that is protesting their nation's policy in Vietnam."

The *NYRB* has already been a constant and invaluable source, both for documenting the perceived parallels between the Algerian and Vietnam wars and to illustrate the theory of cycles of engagement. Its impact and importance cannot be underrated. Few would question Charles Kadushin's conclusion, based on impressive research, that the *NYRB*, at least in the late 1960s at the height of the Vietnam War, was the most influential intellectual journal in this country, "the intellectuals' leading publication," with the most power to make or break reputations.[38] The *NYRB* was notorious enough for former Vice-President Spiro T. Agnew to select it as a special target during his famous attack on the "effete intellectual snobs." In a speech in October 1970 Agnew devised a tongue-in-cheek check list of ten questions "designed to weed out the secret effete." His sixth question was "Does it make you feel warm and snugly protected to read *The New York Review of Books*?"[39]

Six of the ten petitions studied earlier in this chapter circulated in the *NYRB*. Essentially all the important intellectual opponents of the Vietnam War contributed to its pages. Because of its format, which allowed for longer articles, antiwar intellectuals were able to develop their arguments more fully than in the other periodicals to which they contributed. There was a great deal of overlap; for example, an author like Hans Morgenthau also contributed to the *New Republic* and the *New York Times Magazine*. Noam Chomsky wrote for *Ramparts* and served for serveral years as a consulting editor of that passionately engaged, probably *embrigadé* magazine. Andrew Kopkind, who wrote major articles on Vietnam-related topics for the *NYRB*, was contributing editor of the *New Republic* in 1967 and also wrote for and served on the editorial staff of *Ramparts*. Jean Lacouture and Bernard Fall contributed important articles to the *NYRB* and *Ramparts*. Examples of these interrelationships could be multiplied.

We shall now scrutinize more closely the *NYRB*'s unparalleled representation of the cycle of engagement during the Vietnam era.

Engagé *or* Embrigadé?

Many of the 262 articles on the Vietnam War that appeared in the *NYRB* between 1964 and 1975 were of high intellectual and literary quality, and a significant percentage of them were later expanded and published as books. Most of Noam Chomsky's and Mary McCarthy's writings on Vietnam fall into this category. There was a wide diversity of material, ranging from the classic book review, to poetry, to long and sometimes bitter exchanges of correspondence, to excerpts from travel journals (to Hanoi, for example), to extended essays on a general topic relating to Vietnam. Many of these were brilliantly illustrated with caricatures by David Levine.

These writings were indisputably *engagé* but, arguably at least, rarely *embrigadé*. Like the *Esprit* group during the Algerian War, the *NYRB* intellectuals attempted to avoid tunnel vision. Whether or not they wholly succeeded is debatable and probably not susceptible to a convincing demonstration. It is important to set the historical record straight, to show that they at least made the effort, and to give credit where it is due. The *NYRB* has received its share of bad press for its activism during the Vietnam era, in good measure because of its best-known and most provocative cover, that of the August

24, 1967, issue, which included a drawing of a "Molotov Cock-tail" and instructions for its manufacture.[40]

Although they are significantly outnumbered by the wide variety of anti–Vietnam War petitions, three formal petitions denouncing repression in the Soviet Union and its satellites were published in the *NYRB* during the Vietnam years. Per-haps the most telling of these was a March 1968 open letter to Soviet Premier Leonid Brezhnev protesting the recent trial and jailing of four young dissident Soviet writers. It was signed by thirty-five leaders of the antiwar movement, including four of the famous "Boston Five" defendants in the "Dr. Spock Trial," indicted for urging young men to refuse to fight in Vietnam. Other signers included Noam Chomsky, Rob-ert Lowell, Denise Levertov, Dwight Macdonald, and Susan Sontag.[41]

The *NYRB* also published several individual statements by eminent antiwar intellectuals sharply denouncing various forms of repression in Soviet and Czech society, including a powerful attack on the "Inquisition in Czechoslovakia" by Hans Morgenthau, one of the most prolific and eloquent in-tellectual opponents of the Vietnam War. Morgenthau as-serted that the suppression of intellectual freedom by the revitalized Stalinist authorities, in the wake of the briefly lib-erating "Prague Spring" of 1968, was a "scheme of moral emasculation and spiritual destruction." Under the new sys-tem there was no escape for the intellectual short of suicide; and the liar, the informer, had become "the ideal man." In this variant of thought control, men were forced to make their own prisons, and Morgenthau went as far as to characterize it as a "crime against humanity."[42]

In February 1968 Mary McCarthy, who was about to de-part for Hanoi and who had been accused by Diana Trilling of abandoning her critical sensibilities and moving from "in-tellectual to moralizer," commented in the *NYRB* on the shameful behavior of communist intellectuals in the 1950s under Stalin. But since the Twentieth Party Congress of 1956 a new generation of Soviet intellectuals "has been trying to redeem that infamy by courageous speech and action." In fact, McCarthy respected and greatly admired the Soviet dissident movement, contrasting it favorably with the Vietnam protesters at home, who were "too easily discouraged by lack of agreement from Johnson, Rusk, and the opinion pollsters."[43]

Noam Chomsky angrily retorted to the conservative journalist Joseph Alsop, who had accused him and the *NYRB* intellectuals in general of finding all evil in the United States and all good in the Soviet Union. Chomsky reminded Alsop that in the January 2, 1969, issue of the *NYRB*, he had written that "in the grim atmosphere of the Soviet Union, resistance can barely be contemplated. All the more, then, must we honor those who make their voices heard." Chomsky continued with a list of Soviet dissidents whom he praised highly.[44]

There is even in the *NYRB* a powerful attack by Anthony Lewis on the use of torture by the North Vietnamese. The Hanoi regime was usually treated very carefully and mostly favorably by American antiwar intellectuals, especially by those who had visited the country and seen the evidence of the terrible destruction wrought by the American bombing. But Lewis reminds his readers that "truth is not divisible" and that all the stories of torture carried out on American prisoners in the North could not have been invented. "The torture of even one person is inadmissible and so is any attempt to dismiss it as insignificant."[45] Lewis's eloquent and forceful statement reminds us of the admirable moral consistency displayed during the Algerian War by the *Esprit* group and especially by Pierre Vidal-Naquet and the Comité Maurice Audin.

Stage 1

As with the writing in *Esprit* on Algeria, that in the *NYRB* on Vietnam may be divided into four broad categories: first, a perceptive instant history of American involvement in Vietnam and, second, much thoughtful analysis of internal American politics as it was influenced or even dominated by the Vietnam crisis. There is also present in the *NYRB* a lived, documentary history of antiwar engagement during the Vietnam era, through the petition campaigns and the writings of the many *engagé* intellectuals who contributed to the journal. Finally, one can find sensitive and often enlightening discussions, centered always on a response to the Vietnam tragedy, of practically every issue relating to engagement that politically and socially concerned intellectuals face in any historical period. As with the case of *Esprit*, we shall emphasize this fourth area but draw where appropriate on the other three.

The earliest articles in the *NYRB* that make reference to Vietnam are in a concerned and inquiring mode, perhaps really not yet *engagé*; they might be termed *prepedagogic*. The first of these appeared in the September 10, 1964, issue, which went to press less than a month after Congress passed the Gulf of Tonkin Resolution and seven years before the disclosures made in the Pentagon Papers became public knowledge. It already expressed doubts about the legitimacy of the grounds cited by the Johnson administration to secure the resolution's passage. The reference, interestingly enough, is found in a review of a book on the Central Intelligence Agency. The reviewer, Karl Meyer, pointed out that the CIA was coming under criticism for employing "dirty tricks." Given the role believed played by the agency in Guatemala and Cuba, some already suspected that deception was involved, "when the destroyer Maddox was fired upon in the Gulf of Tonkin." Meyer accepted the claim, later challenged, that the *Maddox* was actually attacked but thought that the *Maddox* might have been shelled inadvertently, because the North Vietnamese believed that it was protecting small boats that were landing CIA-trained guerrillas in the North. Meyer contended that the CIA should stick to gathering intelligence and not meddle in the affairs of other nations. He viewed the *Maddox* dispute as probably another example of CIA political blundering but did not predict escalation to the level of a major crisis and war.[46]

The second reference to Vietnam in the *NYRB* is in an article by Paul Goodman analyzing the 1964 presidential election, published in the December 3, 1964, issue. Goodman was worried because many supporters of the Democratic ticket held back criticisms of Johnson's Vietnam policy during the campaign, largely out of fear of what they believed Senator Barry Goldwater might do if elected. A few weeks after the Democratic landslide, we all were so relieved, Goodman reminded us, that we have simply put Vietnam out of our minds. Goodman was, however, decidedly uneasy and wondered whether the military disengagement that Johnson and Humphrey had vaguely promised during the campaign would materialize. Goodman foresaw serious problems, concluding on a foreboding and Orwellian note: "We are rather steadily heading toward 1984."[47]

Exactly ten years after the publication in *Esprit* of Jean-Marie Domenach's "Is It War in North Africa?" and in the very same issue of the *NYRB* that published both the text of Jean-Paul Sartre's statement refusing the Nobel Prize and the D. A. N. Jones article, cited in Chapter 1 as an early example of the Algeria–Vietnam parallels, there also appeared an amazingly prescient article on Vietnam by I. F. Stone, aptly entitled "The Wrong War." No one illustrates better the *pedagogic* level of engagement, the systematic and rationally presented attempt to persuade the government that it had embarked on the wrong course and should change direction, than the wonderfully iconoclastic Stone, a man of the Enlightenment, a believer in reason, an heir to Tom Paine, a true gadfly in the Socratic sense. He had a memory that a historian can only envy, a sharp wit, and, in the words of the historian Henry Steele Commager, an "old-fashioned passion for justice, de-

cency, integrity and honor."[48] It is perhaps not surprising that
in his active retirement in the 1980s Stone (1907–89) taught
himself ancient Greek and devoted years of study and a book
to the trial of Socrates.

Stone wrote a great deal on the subject of Vietnam, short
news reports and press summaries in his own publication,
I. F. Stone's Weekly, and longer analytic articles mostly in the
NYRB, later collected and published in book form. After December 1971, when ill health obliged him to cease publishing
the *Weekly*, essentially all his writing on Vietnam appeared
first in the *NYRB*. Stone understood the issues with extraordinary clarity, foreseeing in his December 1964 article that
Indochina was the region where "the new Johnson administration may most easily stumble into full-scale war after pledg-
ing itself in the election campaign to peace." With some
bitterness Stone pointed out that American policy toward
Vietnam has become exactly that of "an air force Brigadier
General from Arizona." (Barry Goldwater held that rank in
the reserves.)

Believing that the situation had not yet gone too far and
that the United States could still negotiate and save face, at
the end of 1964 Stone thought a settlement was at least possible. We were already napalming villages for "freedom" while
refusing the elections that had been first promised for 1956,
but delayed because President Dwight Eisenhower had been
advised that Ho Chi Minh would garner 80 percent of the
vote. Stone predicted that if we did intervene heavily in Indochina, it would cost as much in lives and treasure as had
France's Vietnam war. His conclusion, based as always on keen
observation and a careful review of public documents—Stone
rarely if ever relied on informants—is uncannily ominous. He
believed that in the Pentagon, the State Department, and the
CIA there were those who advocated a solution of negotiated
neutralization of Vietnam. But, worries Stone, "another, perhaps more powerful group, would rather widen the war than
recognize that it is lost."[49] It is to be regretted that the latter
group did prove more powerful and that President Johnson

did not have the wisdom and the common sense to heed the former and the advice of I. F. Stone.

In April 1965, Stone tried again, in an article entitled "Vietnam: An Exercise in Self-Delusion," ostensibly a review of two Pulitzer Prize–winning books on Vietnam.[50] But as was his wont Stone quickly took off in his own directions. He was more and more distressed about the expanding war, stating that the bureaucrats involved had tried less to deceive the public than to deceive themselves. Both Saigon and Washington authorities had refused to "hear the truth from their subordinates in the field" and were subject to an incredible blindness, a "persistent Panglossianism." When journalists like David Halberstam tried to tell the truth about early defeats in Vietnam, they were viewed as the enemy. The admirals and generals became angry at the press and, when embarrassing questions were asked, replied, "Why don't you get on the team?"

Stone was one of the many American antiwar intellectuals to draw parallels with the Algerian War, commenting on the delusions of the French military elite like Colonels Lacheroy and Trinquier, who thought they could apply "Communist ideas in reverse to the 'pacification' of Algeria." Stone goes on to criticize the Pentagon for overlooking the social and economic roots that produce wars of liberation: The counterinsurgency methods currently used in Vietnam are not only likely to fail militarily as they did in Algeria but also will prevent "debate on peaceful alternatives." There is a growing tendency in the Johnson administration to see any questioning of the judgment of the president as "disloyal," and this reinforces the Pentagon's natural tendency to view as "unpatriotic" any doubts about the validity of resorting to military force in Vietnam. Thus there is a danger of a new McCarthyism as the executive branch and the military together "move toward wider war rather than admit earlier mistakes."[51]

This article by Stone aroused so much interest that it was reprinted, and the editors of the *NYRB* offered to distribute it without charge. Group mailings of over two hundred copies

were *gratis*, otherwise a fee of three cents a copy was charged
and postage paid. Sadly, those who most needed to read it
did not or ignored its message.

In early 1966 Professor John K. Fairbank, for many years
Harvard's leading sinologist, wrote in the *NYRB* that "the
Vietnam debate reflects our intellectual unpreparedness."[52]
By implication, at least once preparedness is attained, political
and military leaders might begin to act reasonably.

With his celebrated four-page *Weekly*, faithfully published
and distributed in fifty thousand copies by himself and his
wife alone, and in the twenty-six thorough articles dealing
with Vietnam that he contributed to the *NYRB*, I. F. Stone as
much as anyone attempted to remedy, for his educated and
influential readership, the intellectual lack identified by Fairbank. Whether he was analyzing the 1968 Senate Foreign
Relations Committee hearings on the Gulf of Tonkin incidents
of 1964; appealing in 1969 to Averell Harriman, who in October 1968, when Johnson ordered a bombing pause, was the
chief negotiator in Paris; or dissecting Nixon's peace plan in
the spring of 1972,[53] Stone was patient, astute, superbly informed, and amazingly perceptive and accurate in his observations and predictions. One might characterize this college
dropout as the perfect pedagogue. The moral tone of course
becomes stronger in his articles as the war drags on, and at
the very end, at the time of the terrible (for both sides) Christmas bombing offensive of 1972, Stone denounced Nixon as
a "moral monster."[54] But he never disclosed any private despair he may have felt, never abandoned the public conviction
that his carefully documented rational argument would have
an effect.

In making this pedagogic effort Stone was by no means
alone among the stable of *NYRB* writers. The eminent French
journalist and biographer of General de Gaulle, Jean Lacouture, in one of his several contributions to the *NYRB*, argued
in March 1966 that a close examination of the "disastrous

errors" made by his countrymen in Indochina between 1945 and 1956, when the last French army units left Vietnam, would be "extremely valuable to American leaders." (Once again, as with Stone and Fairbank, there is the unspoken assumption that our leaders are rational and can learn lessons from the past failures of others.) Through this historical inquiry, Lacouture thought, the Americans should be able to discover "something of what has gone so dreadfully wrong in Vietnam today." The Americans have done no better than the French in identifying and bringing forward viable non-communist leadership. The mandarins are gone, replaced by juntas of young "generals-of-fortune who add a new star to their shoulders after each defeat in battle." Two years before the Tet offensive and the actual brief occupation of part of the U.S. embassy by a Vietcong squad, Lacouture noted that if Hanoi itself were bombed, "we may be sure that the Vietcong forces have well-laid plans to take atrocious vengeance on Saigon, a city they have both infiltrated and surrounded."[55]

Returning once more to I. F. Stone, in a devastating 1966 review of Air Force General Curtis LeMay's autobiography, Stone seized the opportunity to discuss Vietnam. LeMay, a well-known advocate of strategic bombardment was, Stone argued, proved wrong yet again, in that the bombing that began in 1965 "has galvanized North Vietnam into greater and more open aid to the southern insurgents." Seven years before the last wave of saturation bombing of North Vietnam, Stone saw that indiscriminate area bombing simply increased civilian bitterness and facilitated Vietcong recruitment in the South.

Stone contended that one of the important factors that had propelled the United States into the Vietnam War was an internal debate in the command structure of the U.S. Air Force, between those like LeMay who favored bombers and those who advocated missiles. Manned bombers were becoming obsolete, but Vietnam offered the B-52 its "last murderous gasp," and the aim of the massive air force bombing effort in

Vietnam was to show that with manned aircraft we can "transform the war we can't win to a war we might."[56]

Stage 2

Whether or not Stone's hypothesis regarding the hidden motivation behind the bombing strategy used in Vietnam is correct, no one questions that the United States Air Force participated heavily in the effort to effect the transformation he describes. And in so doing it and the other branches of the American armed services employed, as is well known, methods of warfare that many Vietnam War protesters deemed criminal. Beginning as early as 1965 the moral dimension of Vietnam protest is found in a large number of the articles, manifestos, and exchanges of correspondence published in the *NYRB*.[57]

The second level of engagement is superbly illustrated in Robert Lowell's refusal to participate in the June 1965 White House Festival of the Arts. Even though Lowell's letter to President Johnson, published in the *New York Times* and excerpted in the *NYRB*, was extremely polite, it caused quite a scandal, as Lowell was widely regarded as America's greatest living poet. Lowell wrote,

> I know it is hard for the responsible man to act; it is also painful for the private and irresolute man to dare criticism. At this anguished, delicate, and perhaps determining moment, I feel I am serving you and our country best by not taking part in the White House Festival of the Arts.[58]

Twenty well-known intellectuals, including six Pulitzer Prize–winning poets, sent a follow-up telegram to President Johnson supporting Lowell's decision. Although Lowell spoke for himself, "we should like you to know that others of us share his dismay at recent American foreign policy decisions."[59] The incident of the White House Festival, which sharply aggravated President Johnson's antipathy and scorn for intellectuals—and by the end of 1965 the feeling was implacably mutual—in itself merits a separate historical essay. As Dwight Macdonald put it in his acute insider's reportage published in the *NYRB*, "Rarely has one person's statement of his moral unease about his government's policies had such public resonance."[60]

In November 1965 a group of moderate opponents of the Vietnam War, led by Irving Howe, published an analysis of Vietnam protest in the *NYRB*, in which they tried to draw a distinction between "individual moral objection," which they favored, and a full-scale "political protest movement," which they did not. They were reluctant to advocate any form of what we have termed counterlegal action, even civil disobedience, which they believed should be used only as a last resort. In their view, the way to end our morally and politically disastrous policy in Vietnam was "to employ every channel of democratic pressure and persuasion."[61]

None of these methods gave any indication of working during 1966 as the war steadily escalated. Writing for the *NYRB* at the end of that year, Murray Kempton regretfully concluded that a crucial fact about the New Left and also those who are older and view themselves as on the Left and "still have trouble lying to ourselves is that none of us know what to do." Kempton drew a historical analogy with the early twentieth century when the Russian revolutionary movement appeared to be in a state of collapse. We are at a similar crucial moment, when "someone [the reference is clearly to Lenin, who wrote "What Is to Be Done?" in 1902] has to sit down and write across the title page 'What Is to Be Done?' Let us pray the results turn out better."[62]

Stage 3: "The Responsibility of Intellectuals"

Just as the "Call to Resist Illegitimate Authority" is almost definitely the single most important petition to be circulated during the Vietnam years, Noam Chomsky's "The Responsibility of Intellectuals" has a strong claim of being the single most influential piece of antiwar literature. Though this essay was published in a number of places, it is commonly associated with the *NYRB*, where it received its first wide distribution in February 1967.[63] One may view Chomsky's essay as a fragment of an answer to Murray Kempton's anguished request.

The arguments in this famous text were presented in a clear and straightforward manner. Chomsky possessed inexhaustible energy and managed the remarkable feat of continuing to publish scholarly work in linguistics that gained him an international reputation while acquiring a command of a vast literature on Vietnam and related matters. Although "The Responsibility of Intellectuals" was both pedagogic and informed by moral passion, its goal at the time of its publication was to lead its readers to a position in which counter-legal action seemed imperative, almost inevitable. We shall briefly review here how Chomsky arrived at that conclusion.

Chomsky began by emphasizing strongly the privileged position occupied by the intellectuals in Western democratic societies. "They have the power that comes from political liberty, from access to information and freedom of expression." Intellectuals possess the leisure and the professional training to discover truths about contemporary history that lie "hidden behind the veil of distortion and misrepresentation, ideology, and class interest." Because of the benefits that intellectuals enjoy, they have, according to Chomsky, a special responsibility to the societies that have nurtured them, a responsibility that is greater than that incumbent on the general population of a nation. This argument seemed persuasive to many intellectuals at the time and helped gain Chomsky a wide following, and it is arguable at least that it has not lost its validity.

Once he had led his readers this far, Chomsky went on to define the responsibility of intellectuals very simply, focusing on the first of Albert Camus's twin engagements discussed in Chapter 3, "refusal to lie about what we know." In what is probably his single most famous statement, Chomsky wrote: "It is the responsibility of intellectuals to speak the truth and to expose lies."[64] One assumes that in Chomsky's mind Camus's second goal, "resistance to oppression," inevitably flowed from the first.

The problem, according to Chomsky, was that many modern intellectuals declined to fulfill this responsibility, and Chomsky cited as examples Martin Heidegger and Arthur Schlesinger, Jr., in the same paragraph, the former because of his fudging of the truth in a pro-Hitler declaration in 1933 and the latter because of his public admission that he lied to the press at the time of the abortive Bay of Pigs invasion of Cuba in the spring of 1961, when he was part of President John F. Kennedy's brain trust.

The main thrust of Chomsky's essay was a carefully researched, heavily footnoted, and passionately argued attack on establishment intellectuals like Schlesinger, Walt Whitman Rostow, McGeorge Bundy, Henry Kissinger (two years before he became President Nixon's national security adviser), and others, for their abandonment of the critical role that should fall to them because of their privileged status. These academics who had entered politics displayed a kind of "hypocritical moralism" that hid American imperialism with pieties and masked and apologized for aggression, especially American aggression in Vietnam.

Chomsky had a gift for selecting the choice and devastating citation, such as General de Gaulle's critique of the American "will to power, cloaking itself in idealism."[65]

Then there was Daniel N. Roue, the Yale specialist in international relations who advocated in testimony before Congress that the United States buy up Canadian and Australian wheat in order to produce mass starvation in China, and Reverend R. J. de Jaegher, the Seton Hall scholar in East Asian studies who explained that "like all people who have lived under Communism, 'The North Vietnamese would be perfectly happy to be bombed to be free.' "[66]

Chomsky scorned such "scholar-experts," whose professed expertise he dismissed as "self-serving" and "fraudulent." Relying on arguments like those of Daniel Bell, whose influential essay "The End of Ideology" Chomsky also sharply criticized, these new intellectuals were playing an increasingly important role in "running the welfare state." They were in the process of replacing an earlier type of "free-floating" intellectual, who was perhaps synonymous with Julien Benda's *clerc* before he betrayed, and had constructed a proclaimed "value-free technology" to resolve (or pretend to resolve) contemporary social problems. "If it is the responsibility of the intellectual to insist upon the truth, it is also his duty to see events in their historical perspective."[67] In a scathing critique of Secretary of State Dean Rusk, who liked to draw parallels with the Munich crisis to show that North Vietnamese "aggression" must be resisted, Chomsky agreed that the historical parallel had validity but turned the tables, contending that we were aggressor in Vietnam. He pointed out that at the time the mainland Chinese were actively criticizing the Soviet Union, which they claimed was playing the role of Neville Chamberlain to America's Hitler in Vietnam. Of course, Chomsky admitted, "the aggressiveness of liberal imperialism is not that of Nazi Germany, though the distinction may seem rather academic to a Vietnamese peasant who is being gassed or incinerated."[68]

Chomsky concluded his essay with a reference to Nazi war crimes. We must, he claimed, understand the viciousness hidden behind the pseudo-objectively professed by the intellectuals, mostly social scientists, who worked in and on Vietnam in service of our government. We must confront these attitudes openly, "or we will find our government leading us toward a 'final solution' in Vietnam, and in the many Vietnams that inevitably lie ahead." He cited the specific case of a Nazi death camp official who could not understand why the Russians were going to hang him and asked: "Why should they?" "What have I done?" The latter question was one that we should ask ourselves "as we read, each day, of fresh atrocities in Vietnam."[69]

Fulfilling Responsibilities

Hence, one of the ways to answer Murray Kempton's question "What is to be done?" is to ask "What have I done to fulfill the responsibility of intellectuals?" Chomsky's essay and the debate it generated, which brought into play parallels with the intellectual resistance to the Algerian war,[70] may be viewed as a necessary prelude to the "Call to Resist Illegitimate Authority" on October 1967. Relying, as we have seen, on a French precedent rather than a Russian one, the latter text responded in greater detail, making a principled and historically rooted defense of counterlegal engagement.

In a December 1967 *NYRB* article, "On Resistance," Chomsky himself analyzed with force and clarity the question of counterlegal engagement. The crux of the article is a discussion of the October 1967 demonstrations at the Pentagon. Chomsky admitted an "instinctive distaste for activism" but felt himself and others "edging toward an unwanted but almost inevitable crisis." During the march he himself had been arrested, though not ill treated, and he spoke admiringly of the young demonstrators who took the worst of the punishment. "It is difficult for me to see how anyone can refuse to engage himself, in some way, in the plight of these young men."[71]

Chomsky remained firm in advocating nonviolent resistance. Among types of nonviolent civil disobedience then available were draft counseling, which lay in a hazy border area of quasi legality, depending on what was being counseled, and assisting those who wished to escape the country, which was definitely counterlegal. Resembling Jean-Marie Domenach rather closely, Chomsky argued with passion, and I believe persuasiveness, that intellectuals must not recklessly use their eloquence and rhetorical skills to force others—especially the young who are bound to suffer for it much more severely—to commit civil disobedience. "Resistance must be freely undertaken."[72]

"A Time of Engagement"

"The Responsibility of Intellectuals" was published almost six years before the end of formal American military involvement in Vietnam. When the "Call to Resist Illegitimate Authority" and "On Resistance" appeared, the cease-fire was still more than five years away. During that long and painful interlude there was much discouragement among the American antiwar intelligentsia, which is well documented.[73] As Leo Marx observed in November 1967, every issue of the *NYRB* "seems to register a deeper sense of the horror, anger, and frustration felt by a large part of the intellectual community in the face of the Vietnam war."[74]

Yet there was no general retreat into the ivory tower, no widespread abandonment of efforts to end the war or the cultivation of the despair that, as Jean-Marie Domenach had written in 1960, "is the secret weapon of tyranny." Leo Marx himself did not give up, arguing that even if the painful question that George Steiner put to Noam Chomsky in March 1967, "What shall we do?" had not been satisfactorily answered, we must keep trying. Marx thought that there was no single correct answer but suggested as one option an organized boycott of government activities, and he himself resigned from a Fulbright scholars committee. He certainly did not want Senator William J. Fulbright, a leading congressional dove who might have some real political impact, to abandon his office. But because most intellectuals had no serious chance of affecting the government directly, Marx hoped to see a "concerted refusal on the part of the intellectual community to associate itself with this country's reckless foreign policy."[75] In an *NYRB* report on the New Politics Convention held in Chicago in September 1967, Andrew Kopkind portrayed as well as anyone both this mood of frustration and the refusal to succumb to it. To be white and a radical in the summer of 1967 was to "watch the war grow and know no way to stop it." From a strictly logical point of view it should have been a summer of despair and escape, but instead it was "a time of engagement, not withdrawal." Kopkind does not address the question of why engagement persisted, but he does document the tremendous variety of antiwar activities that had sprung up over that summer—suburban housewives canvassing, the development of student draft counseling services nation-wide—while "professors plot demonstrations of protest and non-cooperation."[76] Here Kopkind was referring to the October 1967 mobilization in Washington, which included the March on the Pentagon, and was planned at this time.

A year after "The Responsibility of Intellectuals" appeared in the *NYRB*, Noam Chomsky provided a rationale for continued engagement, at a moment when pessimism and a return to the ivory tower seemed to be rational choices for antiwar intellectuals. His vehicle was an open letter replying to an angry and impatient black militant who, Chomsky felt, "underestimated the force our government commands." After three years of teach-ins and so many other activities designed to raise the public consciousness about Vietnam, it was remarkable, Chomsky noted, that in our democracy there was not one public figure or branch of the mass media that advocated the position that, according to polls, was over-whelmingly supported by our own allies, namely, "that the United States should withdraw from Vietnam."[77] At that his-torical moment Chomsky believed that "by any objective stan-dard" America was the most aggressive power in the world and a true threat to peace. (The truth or falsity of this belief is not at issue here.)

Yet simultaneously, Chomsky admitted, Americans en-joyed substantial individual freedom. It was possible to speak out and organize resistance to the war. Resisters might be harassed and occasionally tried and imprisoned, but there had not been, and one could not foresee, mass executions or in-ternments in concentration camps. Hence even if one were not an uncommonly heroic individual, "resistance is feasible."

According to Chomsky these were the realities, paradoxical as they might appear from the vantage point of other
historical contexts, which one was obliged to face in early 1968
in trying to determine "an appropriate mode of political action." Chomsky was not dogmatic regarding which tactics were
most suitable, instead advocating a spectrum including draft-
card turn-ins. Through a wide range of engagements Chomsky hoped it would be demonstrated "that American ideals
are more than rhetoric, and that American society is too
healthy to permit this war of aggression to continue."[78]

For the duration of the war Chomsky's position remained
firmly counterlegal, though he never went underground,
never went as far as a Francis Jeanson. Even after observing
at first hand in the spring of 1971 the violent police repression
of the "Mayday" demonstrations in Washington, which briefly
turned the nation's capital into a "simulated Saigon," Chomsky
persisted in his advocacy of nonviolent civil disobedience.
Draft and tax resistance should continue; they had served to
focus attention on the war and would help prevent "the cool-
ing of America" while the partially "Vietnamized" war still
raged, with 100 tons of American bombs falling on the region
every hour.[79]

"A Lopsided Category"

Attempts by the American intelligentsia to demonstrate that
their society was indeed, as Chomsky wrote in 1968, "too
healthy" to allow the Vietnam war to continue had begun as
early as 1964, and we have looked at some of their variety
and diversity, such as the petition campaign.

A principal reason for this persistent struggle, despite ex-
haustion and a kind of monotony—the French intellectuals
protesting torture in Algeria a decade earlier recognized that
the shock effect of a litany of horrors eventually wears off—
was that for many intellectuals the Vietnam episode lay in a
special category. It stood outside the normal realm of debate,

the give and take and weighing of both sides that is central to the ordinary life of academics. As Martin Bernal put it in the *NYRB*, regarding America's role in Vietnam there was no obligation to take a balanced view, to look for two sides to every question. For him Vietnam belonged to a "lopsided category, like Nazi concentration camps. A balancing act is not necessary."[80]

By the spring of 1968 many intellectuals had come to believe, as did Professors Gerald Berreman and Frederick Crews, that there was "no longer any middle course between resistance and complicity."[81] Berreman and Crews proposed in a communication to the *NYRB* that because graduate student deferments had just been taken away, professors could put themselves at some risk and resist by signing pledges supporting draft-eligible Americans who refused induction. They should also consider holding separate "Vietnam commencements."[82] In fact, in many colleges and universities around the nation, antiwar professors and students played (seriously) with supposedly sacred spring academic rituals—which normally carry such a freight of symbolism for the young about to embark into the "real" world—by running parallel ceremonies. They also disrupted official programs by wearing peace symbols on their caps and other methods. Vietnam commencements would merit a separate study as part of the history of American higher education.

Varieties of Engagement, 1968–1973

Vietnam commencements, which were popular between 1968 and 1972, were just one of an extraordinary array of ingenious and imaginative engagements, both legal and counterlegal, devised by antiwar intellectuals during the five remaining years of direct American military involvement in Vietnam. The range is staggering, from a "peace evening" in 1971, similar to the "read-ins" that began five years earlier, to specific campaigns on behalf of individual groups like imprisoned

Buddhist monks, to reports by intellectuals who traveled to Paris, met with the official Vietcong and North Vietnamese negotiators, and publicized formal compromise peace proposals.[83] There is no better place to sample that diversity and the ongoing debate over what the responsibility of intellectuals ought to be when confronted with the reality of the continuing Vietnam War than the *NYRB*.

One finds in the *NYRB* reviews of all the important protest literature and of books on the draft and its opposition, which in many ways related to the intellectual life of the nation, especially after the ending of graduate student deferments in spring of 1968.[84] Elizabeth Hardwick submitted a brilliant report on the upheavals in Chicago and the police brutality during the Democratic National Convention late that summer. People kept telling her in pain and bewilderment that they had been lifelong Democrats: "Few had realized until Chicago how great a ruin Johnson and his war in Vietnam had brought down upon our country."[85]

As the 1968 elections approached and Richard Nixon's victory appeared more probable and as hopes for peace dimmed, Margot Hentoff looked back despairingly to the euphoria of the spring, when it seemed as though the peace movement had brought down a government. "It was a time in which anything could happen. It now looks as if everything will." In the wake of the repressive violence in Chicago, "the ideas ran out."[86]

But the *NYRB* documents that this discouragement and intellectual infertility were not permanent, and a seemingly inexhaustible series of proposals in the search for some impact on the goverment's Vietnam policy appeared in its pages. Some of them were perhaps fanciful, such as Arthur Waskow's October 1968 idea of voting no for the president in the regular November election and setting up alternative "freedom elections."[87] Others like tax refusal, a frequent topic, were probably effective, in that the government had to divert labor and resources to recoup these funds at a certainly higher cost than the value of the sums recovered.

It is generally agreed that mass-based opposition to the Vietnam War reached its high-water mark in the fall of 1969,[88] although there was no apparent weakening of antiwar resolve among the intelligentsia. In the *NYRB*, analysis of the war and its historical roots continued.[89] There was also extensive discussion of new phases of the antiwar moment, such as the "ultraresistance" represented by Father Daniel Berrigan and the Catholic Left. Berrigan, one of the "Catonsville nine," whose trial transcripts he transformed into a remarkable play, was now firmly counterlegal.[90] His small group of Catholic militants, who believed that they were representing the true spirit of the Gospels, were convinced that they had exhausted all legal means of anti-Vietnam protest.[91] Their activities were carefully tracked in the *NYRB* throughout the remainder of the Vietnam War,[92] and Daniel Berrigan's marvelous series of letters and dialogues from the "Underground," composed while he was in hiding in the summer of 1970, were first published there.[93] This is yet further evidence of the centrality of the *NYRB* in studying anti-Vietnam engagement. Tom Wicker could have not known it three years earlier, but his choice of the metaphor of the Bible was appropriate in more ways than one! Father Berrigan was finally caught by the FBI in August and served his sentence in the Danbury, Connecticut, federal penitentiary.

Father Berrigan's writings from this period may be viewed as a quintessential testament of advocacy for what we have called the counterlegal position on intellectual engagement. As Berrigan put it in describing the burning of draft records with homemade napalm in Catonsville, Maryland, on October 27, 1968, "I never tried to hurt a person. I tried to do something symbolic with pieces of paper."[94] Basing his arguments on his own unique blend of theology, history, and poetry,

Berrigan defended his trip to Hanoi, which had deeply moved him while profoundly offending many Catholics. He also defended his decision to go underground after he and his eight coconspirators were tried, found guilty, and sentenced for their "symbolic" actions in Catonsville. Some of his arguments are clearly relevant to secular engaged intellectuals such as Noam Chomsky and Susan Sontag, both of whom also traveled to Hanoi, and indeed all three names frequently appeared on the same petitions. There are times, Berrigan contended, in an elegant reformulation of the responsibility of intellectuals, when great pressures are placed on all professions, not just the clergy, to question the established ways of doing things, "and we each of us must move our professional life to the edge, so to speak, and begin again from the point of view of a shared jeopardy."[95] He called for a "profound form of nonviolent humanitarianism, or we become part of the proliferating problem."[96]

The fall 1969 "mobilization campaign," part of a mostly nonviolent and legal ongoing pedagogic effort, novel in its systematic and nationally organized aspects, was well covered in the *NYRB*.[97] There were reviews of show trials mounted by the government in its ongoing effort to suppress, or at least mute dissent, including those of the "Boston Five" and the "Oakland Seven."[98] Henry Kissinger, who became the subject of numerous articles in the *NYRB*, is quite judiciously evaluated by Clayton Fritchey, and a perhaps grudging tribute is paid to his shrewdness. President Nixon and his national security adviser were viewed as much more adroit than Johnson was, as they proceeded with just the right balance of troop withdrawals to satisfy some moderate war critics, while keeping Thieu in power. But Fritchey did suspect in September 1969 that no matter how clever Nixon and Kissinger were, they might not be able to "keep anti-war sentiment from exploding."[99]

*

Fritchey's speculation proved correct in May 1970, after the debacle of the American military incursion into Cambodia, which triggered a first, relatively benign, wave of campus protests. Then the killing of four students by the National Guard at Kent State University on May 4 led to one last resurgence of mass activism nationwide.[100] The *NYRB* documented these tragic events in many ways, including a moving short poem by Paul Goodman who had recently lost his own young son in an accident. Its last two lines are as follows:

> Today they have begun to massacre our
> own children. Call the soldiers back.[101]

In the wake of the disasters of the spring of 1970, faculties such as that of Princeton followed the lead set by Harvard in October 1969 and actually made group engagements, public statements condemning the Cambodian invasion. Though not illegal and probably largely morally inspired, such actions were both novel and significant. Normally faculties as organized bodies have been and remain extremely reluctant to make an institutional commitment to any political position.[102]

As a result of the turmoil that shut down most American colleges and universities for a week or more in May 1970, the so-called Princeton plan was devised to release students from classes for a set time period in the fall of 1970 so that they could campaign for candidates in the congressional elections. This plan was adopted at Vassar and elsewhere. What happened during the Cambodian crisis, as Lawrence Stone wrote in the *NYRB*, was that the actions of the Nixon administration activated at least momentarily "the moderate middle of both student and faculty who had hitherto gone quietly about their business and had left political protest to the activists of the S.D.S."[103]

Occasioned by the publication of Telford Taylor's important and influential *Nuremberg and Vietnam: An American Tragedy*, 1971 brought a full and open debate in the *NYRB* over the appropriateness of applying to Vietnam the precedents believed established by the Nuremberg trials. In some sense this was a vindication for the *NYRB* intellectuals, many of whom had been for several years calling attention to the Nuremberg precedents as they related to the American involvement in Vietnam. Now Taylor, who was an eminent professor of law and had served as the United States' chief counsel at the Nuremberg trials, actually admitted the possibility of some limited applicability to Vietnam of precedents set by war crimes trials. Even if, as Jonathan Mirsky noted in the *NYRB*, Taylor's analysis was not "damning enough," it still was an "important contribution to consciousness-raising."[104]

Another aspect of consciousness raising was the insight into the origins of the Vietnam War provided by the Pentagon papers, which were given to the press by Daniel Ellsberg in 1971 and thoroughly reviewed in the *NYRB*. The doubts that American intelligence experts had from the beginning about the viability of our policies in Vietnam now came out into the open, as did the consistent "policy of lying" that hid those doubts.[105]

During 1972 the hesitant and fitful steps toward peace were carefully and critically analyzed in the *NYRB*;[106] and as the end of the war, or at least the American involvement in it, appeared more and more probable, the issue of amnesty for draft resisters and other opponents of the war was sharply debated.[107] The savage ongoing combats, now fought primarily by Vietnamese themselves, were not ignored.[108] The South Vietnamese regime, despite the 1.1 million people officially in its armed forces and the 200,000-person police force, was barely able to contain a North Vietnamese spring offensive in 1972, and then only with the massive assistance of the U.S. Air Force, which mined Haiphong harbor for the first time on May 8 and attacked rail and road connections to China. As I. F. Stone observed, seven years after he made his first prediction of America's defeat in Vietnam, air power had not given us victory, but "only prolonged the agony before defeat."[109]

"The Defeat of America"

In September 1972 I. F. Stone wrote in the *NYRB* of his fear that the agony of open warfare would be prolonged until 1976, and for the Vietnamese people he proved to be only a year off the mark.[110] As the negotiations in Paris inched toward agreement on a cease-fire that would eliminate direct American involvement in the war, the question of standing "before the bar at Nuremberg," as Philippe Ivernel had so eloquently put it concerning his own country ten years earlier, simply would not disappear. The poet Basil Paquet, who had served as a medic in Vietnam and claimed to have witnessed firsthand the Americans' barbaric treatment of Vietnamese civilians, reviewed with a combination of sensitivity and outrage the issues of war crimes and the Nuremberg principles. He related them specifically to the trial of Captain Ernest Medina, charged with responsibility for the massacre of March 16, 1968, committed at Sonmy (usually known in this country as Mylai). Basil Paquet regretted that with the acquittal of Captain Medina, awareness of American responsibility for war crimes in Vietnam would evaporate from the public consciousness. Paquet commented favorably on a proposal that Senator George McGovern had made (to be sure it was never acted on) that Congress ask a group of foreign nations to test the legality of the Vietnam War under the Nuremberg rules and the United Nations Charter. "That would be one way to achieve fair and just trials for our crimes, and justice for the Vietnamese people. Perhaps Johnson would finally get the chance to serve another term."[111]

A month later, in October 1972, the broad implications of the Nuremberg parallels were reexamined in the *NYRB*, this time not by a brilliant journalist with intellectual qualifications like I. F. Stone, or a militant engaged intellectual like Noam

Chomsky or Basil Paquet, but by the sober, immensely respected academic historian, Henry Steele Commager. Commager, born in 1902, was one of the most famous, prolific,
and widely read historians of his generation. He was a professor at Amherst College from 1956 until his retirement in
1972, received two dozen honorary degrees, and between
1954 and 1964 lectured all over the world including Europe,
Israel, and Latin America under the auspices of the Department of State. It would be hard to imagine a more "establishment" figure. (Interestingly enough, however, after 1964
he was not again invited to lecture on behalf of the Department of State until 1975, when he spoke in Japan.)

Yet Commager had been a firm opponent of America's
military involvement in Vietnam since 1963 and during the
years of open conflict had contributed three major articles on
the war to the *NYRB*.[112] Commager did not feel obliged to
wait for decisive events that appeared imminent but had not
yet occurred and entitled his essay "The Defeat of America."
It was one of the most famous articles to appear in the *NYRB*
during the Vietnam era and was reprinted separately and
widely circulated.

Commager seized the occasion of a review of Richard J.
Barnet's study of the origins of American involvement in Vietnam, *Roots of War*, to suggest that American scholars now
resembled their German colleagues. The former were in 1972
already asking what had obsessed the latter for twenty-five
years, "how to explain the catastrophe." True, the continental
United States had not endured physical damage as did Germany during World War II—only Asians had suffered in this
way—but "morally it was a catastrophe for the American people analogous to that which so profoundly disturbed thoughtful and historical-minded Germans." Of course, Commager
admitted many differences, including that the fact the German conscience could not find expression until after defeat,
which was no help to victims of Nazi terror. And in the United
States, universities "still shelter dissident scholars." There is

a level of awareness and a sensitized conscience on the part of at least some Americans, and this "may be effective in mitigating the ravages of American policies."[113]

Using the documents then available, a close examination of the long and gradual escalation of the war demonstrated the tremendous difficulty in finding any coherent logic or rational objective behind our involvement in Vietnam, beyond some vague "honor" that in any case we have forfeited. For Commager there was a "demented" quality to the American military effort in Vietnam; by 1972 with Nixon's trip to China and the thaw between Washington and Beijing, the original goal, if goal it was, of containing China had obviously been abandoned.

Commager reviewed evidence and arguments from eight years of governmental deception, including the corruption of language. Bombing was "protective reaction"; concentration camps were "pacification centers"; and so on.[114] We Americans had blithely come to accept a double standard: German reprisals in World War II were seen as war crimes punishable by death, "but when we wipe out defenseless villages with 'incontinent ordinance,' or engage in massacres as brutal as that at Lidice, these are mistakes or aberrations that do not mar our record of benevolence."[115]

President Nixon, like his predecessor, was determined not to go down in history as presiding over the first American defeat,[116] but Commager believed it was too late. America had unalterably been defeated, "not, to be sure, on the field of battle, but in the eyes of history." We have the power to destroy Vietnam and call it victory, but such insensate destruction would be madness, not victory. This is a war we cannot win and "a war we must lose if we are to survive morally."[117] Commager concluded that the true patriots throughout this awful episode in American history had been the war resisters.

When the "Agreement on Ending the War and Restoring Peace in Vietnam" was finally signed in Paris in January 1973,

Frances FitzGerald carefully analyzed the text and its implications for the *NYRB* readership, predicting that combat
would continue at some level of intensity while the United
States painfully and slowly pulled out, in a sense replaying in
reverse the film of how it had entered and escalated the war.
Reflecting quite accurately the design of the Vietnam veterans'
memorial that was eventually constructed in Washington,
FitzGerald wrote that "there will be no real end to it, any more
than there was a real beginning."[118]

For more than two years thereafter, every major twist and
turn of the struggle for power in Vietnam, and the impact of
the developing scandal of Watergate on the American presence in Indochina, were analyzed in the *NYRB*, often with
great perspicacity.[119] The reduced American presence in Vietnam did not diminish the concern of *NYRB* writers for the
suffering of the Vietnamese people.[120] Efforts to obtain the
release of Vietnamese political prisoners, many of whom were
still imprisoned under appalling conditions after the January
27, 1973, agreement, were publicized in the *NYRB*.[121] It also
played a leading role in a widespread effort by American
professors and other intellectuals—one of their last Vietnam-
related, politically *engagé* gestures—to nominate and secure
the award of the 1974 Nobel Peace Prize for "all those young
persons who refused complicity with their government's war
in Vietnam—whether through draft refusal, desertion from
or opposition within the armed forces."[122] As is well known
the 1973 prize had been awarded jointly to Henry Kissinger
and Le Duc Tho for their role in negotiating the Paris accords.

NYRB authors took note of the general decline of engagement, which dropped off precipitously after January
1973, though it did not disappear completely until the final
North Vietnamese victory in 1975. During the week of Nixon's
second inaugural, Francine du Plessix Gray lectured at Amherst College, where seven months earlier in May 1972 more
than one third of the faculty and student body had been
arrested in a protest against the Vietnam War—yet another

indication of the widespread and long-lasting nature of the counterlegal engagement that had begun in 1967. Many of the students—for whom the apparently endless war had been a permanent presence since grammar school and the commanding event of their young lives—were now feeling a "strong sense of disorientation" and an uncertainty about what would happen to the "movement" once peace became a reality. It was as if American soldiers had to continue dying to keep the peace movement alive, "as if this cancer was necessary to remind us that we were ill."[123]

"The Meaning of Vietnam"

The last article in the *NYRB* that is technically a primary source for the Vietnam War is entitled "The Meaning of Vietnam," and it was commissioned by the editors just as Saigon fell at the end of April 1975. It consists of reflections by thirteen of our best-known intellectuals on lessons that might be drawn from Hanoi's victory.

All the comments are sensitive, and there is not a self-congratulatory note among them. There is hesitation about even attempting a final statement; as Elizabeth Hardwick observed, "one's adjectival vehemence has been used up."[124] For the historian Christopher Lasch, who had been involved in Vietnam protest for a full decade, our intervention in Indochina was "a grotesque mistake from the start," and he saw no good deriving from the victory. After thirty years of savage struggle, there would be no gentle government in Vietnam. "Only sentimentalists will think that the Vietnamese are now going to enjoy 'democratic socialism, popular rule, and civil liberties.' "[125]

Some of the contributors are wistful, expressing genuine regret that they had not been able to do more to hasten the end of the war. Norman Mailer believed that "the responsibility for the war was entirely ours" and that the effect of the Vietnam war on our national life and our global position had been almost entirely negative.[126] Mary McCarthy thought that the time to withdraw would have been 1968. One could say that time stood still for seven years; since 1968 nothing basically changed "except the color of the corpses."[127]

How the Vietnam episode would and should appear in historical studies was already a concern. Noam Chomsky predicted, correctly, that a battle over the meaning of Vietnam would develop: Even if our government was defeated in Indochina, it was "only bruised at home," and so Chomsky was worried about what would happen to the historical record, "as the custodians of history set to work."[128] The question of memory and history was central also for Sheldon Wolin, who quoted President Gerald Ford as saying that the lessons of the past had been learned and that we should focus on the future. The past, Ford declared, should be left to the historians. Wolin partly accepted this stricture as common sense—it is unhealthy to pick over endlessly the bones of the past, and a "politics of oblivion" might be advisable. Yet he remained ambivalent as to whether such forgetting was desirable or even possible. Rather, the Vietnam horrors are "an indelible part of our history."[129]

Gary Wills saw the Vietnam war as a social thermometer for this country; when it was invisible Americans ignored it and when it could not be hidden it became a kind of civil war, "hawk against dove, bombing Hanoi to impress the Woodstock Nation, a kind of bloodless infanticide, with Asians to do the dying for us." If we want to understand Vietnam, Wills continues, we must see it not as particularly American but, rather, as a relatively minor but artificially inflated episode in the larger historical process of decolonization. France had already lost in Vietnam, and the French were more skillful than we were in disengaging. It is stupid to go over what went wrong. "The only wrong thing was being there."[130] As I. F. Stone had written in the same periodical over ten years earlier, it was "the wrong war," and it was lost from the beginning. A second cycle of engagement could now close.

Eggheads?

When one reflects on these two episodes in intellectual history and on their relation to the "larger" military, political, and social histories in which they are embedded, when one wonders what would have happened had Jean-Marie Domenach been heeded by those in power in December 1954, and I. F. Stone in December 1964, one may question an old American refrain concerning intellectuals. They have often been pictured as naive "eggheads," "terrible simplifiers," idealist pie-in-the-sky types, yet potentially dangerous with dictatorial inclinations if they were ever to be given political power. Perhaps, as our French friends revisit Algeria and we revisit Vietnam and together we consider the needless and senseless horror, death, suffering, destruction, and criminality that occurred in those two lands during the process of decolonization, it is time to reexamine old stereotypes about intellectuals. And it might be advisable to hurry, for if Bernard-Henry Lévy is correct, the species is about to disappear.

5

The End of the Intellectual?
The End of Engagement?

It would be hard to deny that the years between 1954 and 1962 (for France) and between 1964 and 1975 (for the United States) were extremely turbulent, dramatic, and often painful. This would be true even if we set aside factors such as those discussed in Chapter 1—political, diplomatic, and military.

There was yet another ugly side effect of both the Algerian and Vietnam wars, which we have only touched on thus far in this work and might conveniently be labeled *social*. The generally disaggregative effects of the wars on French and American societies produced intense levels of hatred and a sometimes savage civil strife. To describe this and similar periods of extreme internal dissidence in the past, French scholars coined the phrase *guerre franco-Française*, which deserves translation into English to describe remarkably similar phenomena. For the United States during the Vietnam era, one could perhaps speak of the "war within the states," rather than the "war between the states," commonly used by Southerners to refer to the American Civil War. Indeed, George C. Herring believes that the Vietnam War "polarized the American people and poisoned the political atmosphere as no issue since slavery a century before."[1]

To identify the French war in Algeria as concomitantly a *guerre franco-française*, and the American war in Vietnam as concomitantly a "war within the states," seems quite reasonable when one ponders the numbers of assassinations and assassination attempts; the riots; the burnings; the death and devastation caused by bombings in both countries; the activities of the OAS, the Weathermen, and so many other groups; Kent State, Métro Charonne, Chicago 1968, Washington, D.C., and Mayday 1971; and the casualties on all sides, not to mention the destruction of property.

In fact, observers in France have commented that in the end, the parallel *guerre franco-française* of 1954–62 was much more important to the French than was the actual war against the FLN fought on Algerian soil, and it remains so to this day.[2] Or as Gloria Emerson put it, beautifully if bitterly, in 1978: "Each year that it lasted Americans who took opposite sides on the war seemed to hate each other more than the Vietnamese who opposed us."[3] Despite occasional nostalgic looks back to Algérie française in the French media, and some nostalgia in the American media for the excitement of the 1960s, I cannot imagine that any thoughtful person, once reminded of it, would want to resurrect the abysmally divisive internal social situation in the home countries that accompanied the Algerian and Vietnam wars.

If we return our focus to intellectual history and specifically the history of the intelligentsia, few would question the appropriateness of Sandy Vogelgesang's choice of biblical language in the title of her work on the American intellectual left and the Vietnam War. The decade of war was indeed "the long dark night of the soul." The evidence presented in Chapter 3 suggests that the phrase is equally applicable to the majority of French intellectuals during the Algerian War. Certainly it is difficult to imagine how one could make a serious case that it would be a good thing for the intellectual elite of either nation to enter and endure another such period of torment, anguish, and internecine struggle.

One element that we have touched on periodically but not emphasized in studying intellectual engagement is that the

general social disruption caused by the Algerian and Vietnam
wars produced in the French and American intelligentsias very
high levels of confusion, bitterness, and internal feuding, a
dour mood that is difficult to recapture. For many of the surviving intelligentsia the savage and almost entirely fruitless internal polemics during the long years of war are painful to
recall. In an April 1968 debate with Paul Goodman, Robert J.
Lifton used words that could have been written in 1958 in
French by intellectuals as diverse as Jean-Marie Domenach or
Jean-Paul Sartre. One of the casualties of the war in Vietnam,
Lifton argued, was that the rage and frustration of the antiwar
intellectual became directed "not at real enemies but at those
whose purposes are closest to his own." Among opponents of
the war there was an "entropy of protest—a wasting of energy."[4]

Eugene Genovese, like Staughton Lynd, was a historian
by profession and an articulate and committed antiwar militant (in 1965 Lynd was one of the first American opponents
of the war to travel to Hanoi; in 1967 Genovese caused a real
scandal by being probably the first major American intellectual figure to call publicly for the victory of the Vietcong).
Yet in December 1968 Genovese believed that Lynd had been
exposed as a "demagogue," whose philosophy, "notwithstand-
ing his pious speeches on nonviolence, . . . is an incitement to
totalitarian violence."[5] In January 1969 Michael Harrington
thought that Dwight Macdonald was "more reprehensible
than a wiretapper."[6] This is but the tip of the iceberg; examples of this kind of counterproductive verbal violence
among antiwar intellectuals could be multiplied in both
countries.

Yet André Mandouze—a Catholic militant, a professor, an
honored Resistance leader during the Nazi occupation of
France, later an important opponent of the Algerian War,
and a signer of the Manifesto of the 121—could reminisce in
1987 on his prior engagements and conclude: "What a pity
that there is no longer an occasion for the Algerian War." He
admitted that he was being provocative in saying this, but in

seeing how we have all "flattened ourselves out, it is highly evident that we are searching for great causes."[7]

Mandouze may well have been reflecting an unease about the future of intellectuals and the future of engagement that is documented elsewhere. As early as 1977 Paul Sorum asserted that the activities of the French intellectuals during the war in Algeria "may prove to be the final great battle in the long tradition of France's 'engaged' intellectuals."[8] And Sandy Vogelgesang had already suggested in 1974 that the opposition of the American intellectuals to the Vietnam War "may have been a rear-guard exercise in futility against the Age of Technology."[9]

Vogelgesang's formulation implies that other factors—only tangential or at most indirectly linked to foreign war and decolonization— may be permanently separating intellectuals from the engagement that had been a sporadic yet central part of their perceived social role. Perhaps we really are in the "Empire of the Ephemeral," to borrow the title of Gilles Lipovetsky's intriguing and controversial 1987 book praising the virtues of constantly changing fashion.[10] Lipovetsky is a professor of philosophy at the Lycée Emmanuel-Mounier in Grenoble. Mounier (1905–50), the founder of *Esprit* and principal theorist of "engagement" in its modern sense, must, as one reviewer of Lipovetsky's book pointed out, be "turning over in his grave."[11] Are we in truth "changing epochs," as the "new philosopher" Alain Finkielkraut lamented in a succinct critique of Lipovetsky's religion of consumerism and worship of *la mode*? For Finkielkraut, *l'engouement succède è l'engagement* (sudden infatuation has taken the place of engagement)[12]

The "Sartron"

This brings us full circle to another member of the "new philosophers" group, Bernard-Henri Lévy, and his "Elegy for the Intellectuals" of 1987, with which we began. After stating dramatically his hypothesis that along with manifestations of engagement, intellectuals are about to disappear from "this France that invented them, praised them to the skies, dragged them through the mud, but always with passion," Lévy goes on to discuss in a more methodical way some of the reasons he finds for this "*débâcle*."[13]

Lévy, who has often appeared on television and has made effective use of that medium, believes that it is a gross over-simplification to put all the blame on the electronic media and that intellectuals themselves are to a degree responsible for their own "degradation." Part of the explanation lies in the nature of structuralist and poststructuralist philosophy, which according to Lévy has led to a banalization of culture. Another factor was the "demolition" of Marxism, which had been enormously appealing and an incitement to political activism for several generations of French intellectuals. Lévy claims (legitimately) some credit for his "new philosophers" group, which mounted a strong campaign against Marxism in the late 1970s, but worries that nothing has arisen to take its place.

Also important was the unprecedented disappearance, whether through death or, in the case of Louis Althusser, through madness, of almost all France's internationally known intellectuals—from André Malraux who died in 1976 to Simone de Beauvoir in 1986, with Jean-Paul Sartre, Raymond Aron, Roland Barthes, Jacques Lacan, Michel Foucault, and others in between. Their deaths were sudden and in some cases unexpected, and their departure from the scene left a void in French cultural life; no one of their stature has appeared to replace them in the chain of intellectual generations.

Lévy's principal explanation is, however, based on an ingenious acronym, the "Sartron," a blending of the radical existentialist Sartre (1905–80) and the inveterate liberal Aron (1905–83), who after having been close friends in university days became bitter ideological enemies immediately after World War II. For Lévy the key date was June 20, 1979 and a dramatic and moving reunion of the two world-famous in-

June 20, 1979. The genesis of the "Sartron," the reconciliation of Jean-Paul Sartre and Raymond Aron during a meeting of a committee set up to aid the Vietnamese boat people. Between them, the "new philosopher" André Glucksmann. Photograph courtesy of J. R. Roustan and *l'Express.*

tellectuals in Paris at the Hôtel Lutecia, in a meeting in support of the Vietnamese boat people. They met once more, 6 days later, at the Elysée Palace for the same purpose. I cannot emphasize enough the multiple ironies here as they relate to the principal subjects of this book.

Even if Sartre, blind, ill, and near death, and Aron, who had suffered an almost fatal heart attack three years earlier, did not intend their much publicized handshake to take on a vast symbolic meaning, it became "the inescapable model of relations between intellectuals."[14] Instead of sharp debate and disputation, as in the past, a passionate quarrel over the nature of both the ideal and the real society, a strong sense of responsibility for others, indeed a concern for the entire planet, we have a clone, the Sartron, in the Parisian firmament. According to Lévy, the role of the *clerc* in an age of "egotistical retreat" is to reach agreement, and consensus has become a new religion. So in a kind of self-destructive masochism, intellectuals have invalidated themselves. In reducing the tasks of thought, intellectuals have with full consciousness "proclaimed their uselessness."[15] Thus both engagement and the traditional type of intellectual will soon disappear.

Lévy is not totally pessimistic and indulges in some speculation of his own about the future. He thinks that a new third type of "less-engaged" [*moins engagé*] intellectual may be on the horizon. He is convinced that as the twenty-first century approaches there will be less controversy and public spectacle, less passion. Intellectuals, if they exist at all, will adhere only partially to a political or social cause, will keep part of themselves *dégagé*. There will be, he asserts, no more "manifestos of the 121"—such group texts are "out of date."[16] Needless to say, if Lévy is correct, there will be no further "calls to resist illegitimate authority" either, and no Chomsky to define the "responsibility of intellectuals" in the new century, helping propel his peers into a new cycle of engagement.

Lévy's "third type" of intellectual bears some resemblance to H. Stuart Hughes's "sophisticated rebels," the extremely diverse group of individuals, ranging from Pope John Paul II to the German "Green" party leader Otto Schily, treated in Hughes's elegant new study of European dissent between 1968 and 1987.[17] Hughes describes a more cautious and careful kind of involvement, normally focused on a fairly narrowly defined cause, which should probably not be understood as engagement in the traditional sense.

Looking Toward the Year 2000

So, then, is Lévy correct? Will dictionaries a dozen years hence describe *intellectual* as a vanished social type, ranging it with nouns like *serf*, *knight*, and *Centurion*? Will *engagé* disappear from French dictionaries and English usage, and *engagement* refer solely to various contractual arrangements, often connected with marriage?

Some historians, such as Pascal Ory and Jean-François Sirinelli in important work in France, are not so sure. They see intellectual engagement as following a pattern of flux and reflux, beginning with the Dreyfus affair but with a generally ascending curve throughout the twentieth century.[18] They emphasize the importance of the period of the Algerian War as a time of mobilization of the *clercs*, which may be viewed as a "sort of golden age of the intellectual *engagé*." Even though a quarter of a century has elapsed since the 1962 Évian accords, the debate over the role of intellectuals during the Algerian War remains an "affair of conscience and not of historical science, because the wounds have not yet healed."[19]

Ory and Sirinelli published their book in 1986, and they admit to some difficulty in explaining the extended period of disengagement, the "era of emptiness," that has lasted almost twenty years. It began in 1971, when a kind of "lassitude" set in[20] following the intense activism that peaked with the student movement of 1968. They suggest astutely that intellectuals in the 1980s are in a stage of *échaudement*, which might be loosely translated as "burnout." Judging themselves "to have been deceived by one or several previous engagements, the intellectuals refuse any new mobilization."[21] In their view, the intellectual class is not moribund but, rather, in a period of "mutation," with many future possibilities open to it.[22]

Will one of those possibilities be a resurgence of engagement? Even though the thrust of their earlier argument suggests that engagement will recur, Ory and Sirinelli, as historians and not futurologists, are hesitant to make predictions. Surely it would be hard to find evidence that as late as the year of the bicentennial of their great revolution, French intellectuals have made even fitful steps toward entering a new cycle of engagement.

In the United States since the end of the Vietnam War

there has been some fluctuation between complete disengagement and Stage 1, but no movement beyond the pedagogic.[23]

My own view is that Ory and Sirinelli are at least partly correct
and that the intellectual classes in advanced industrial nations
like France and the United States are in a stage of "mutation"
rather than approaching extinction. The intellectuals could
redefine and reenergize themselves, through a resurgence of
activism that would make the curve of engagement rise again
if two factors appear coterminously.

First, a government in power has to do something stupid
and evil enough to elicit a profound moral reaction from its
intellectual elites. Given any acquaintance with the long span
of recorded human history, one may conclude that this is not
an impossibility.

To focus on our more recent case, let us not forget that
for several years, especially between 1967 and 1971, significant numbers of American intellectuals—ranging from graduate students to senior professors, from rabbis, ministers,
nuns, and priests to physicians, artists, novelists, and poets—
were willing to take the third step into full counterlegal engagement. In Chapter 4 we looked at some of the range of
their activities, from helping young men escape to Canada to
traveling to Hanoi, thereby risking popular hatred, to say
nothing of legal action.

Mary McCarthy admitted in a letter to the *NYRB*, shortly
before her departure for North Vietnam, that the "power of
intellectuals, sadly limited, is to persuade, not to provide
against all contingencies. They are not God." And they are
rarely politically gifted. "What we can do, perhaps better than
the next man, is smell a rat." This happened with the war in
Vietnam, "and our problem is to make others smell it, too."[24]

I do not believe that intellectuals have lost their acute ol-
factory sense. And should that sense be reactivated, they
would move quickly to the moral level of engagement, when

Paul Ricoeur's "ethic of distress" would again come into play. In order for significant numbers of intellectuals—united by a shared moral commitment if not by ideological or political affinities—to take the next step and adopt some perhaps new and as yet untried varieties of counterlegal engagement, a second factor must also be present. The external historical situation, the context in which their moral sensibilities have been reignited, must not appear totally hopeless and impermeable to change.

To illustrate this point I turn to Michael Ferber's brilliant statement, "On Being Indicted," of April 1968. At the time Ferber was a twenty-three-year-old graduate student in the Harvard English department and the youngest member of the "Boston Five," who were under indictment and awaiting trial for conspiracy to violate the Selective Service Act. Ferber attempted to explain why so many had joined the new Resistance with the attendant risk of imprisonment:

> Men whose insides were ready for commitment needed only the barest hope of a chance that their gesture would be more than an act of moral witness, that with sufficient numbers and organization they just might have a measurable impact. They were willing to pay a high price; all that was needed was the chance that the price *might not be for nothing*.[25]

It is at least conceivable that these two preconditions for a resurgence of engagement will converge. Hence I would not bet on Bernard-Henri Lévy's dictionary definition for the year 2000, ingenious as it may have been. Perhaps the entry in Webster's will read: Intellectual, noun [from the French, *intellectuel*], a social and cultural category first described in Paris at the moment of the Dreyfus affair and quickly adopted into English. Refers to men and women given to the exercise of the intellect, but also prone to periodic intervention in public life. See Engagement.

EPILOGUE

Questions of Memory

In March 1962, just after the Évian accords, "A. L.," a young veteran of the Algerian War, meditated in Paris on the peace that he wanted and the independence that he felt the Algerians deserved. He sympathized with the Algerians and the festivities that would be soon taking place—the chanting, the dancing, the sacrifice of the lambs. But he could not celebrate with them. "Too many faces remain between peace and myself," those of friends killed, wounded, driven insane by this war. He understood and envied the joy of his Algerian friends and asked them to pardon him. "A. L." remained alone in the French capital with the memories of his friends who had died without fame, without glory. "Pardon me, but there is no forgetting here, we have no victory to celebrate, whatever may be written on the walls of Paris and electoral posters."[1]

A. L.'s age cohort of men born between 1933 and 1941 may have gone through an identity crisis[2] and continue to be unable to exorcise the painful memories of the war in which about three million of them served. Just as around a half-million out of the three million American veterans of Vietnam will suffer some form of PTSD (posttraumatic stress syndrome) during their lifetimes, many French veterans of Algeria will remain locked into the "silence and shame" (*le silence et la honte*) so brilliantly studied by the psychiatrist Bernard W. Sigg in his 1989 book subtitled "Neuroses of the Algerian War."[3]

In the American and French populations at large, however, the situation has been different. For the approximately fifty million French men and women who neither served in the Algerian War nor are North Africans living in France (whether *pieds-noirs*, Arabs, or Berbers), there has been forgetting; there have been quite remarkable "troubles in the French memory."[4] Many writers and observers have described France "as if struck by general amnesia with regard to the Algerian War, as soon as the curtain fell on the tragedy."[5] The Algerian War simply does not have a status in the "French national consciousness and in the national memory."[6]

There is an almost perfect meshing here with the post-Vietnam situation in the United States, where in the words of George C. Herring "in the immediate aftermath of the war, the nation experienced a self-conscious, collective amnesia."[7] As Mike Mansfield, who as the Senate majority leader had for many years been closely involved with Vietnam policy, put it in 1977: "It seems to me the American people want to forget Vietnam and not even remember that it happened."[8] Also in 1977, recalling a fellow officer who had been killed in Vietnam eleven years earlier, Philip Caputo wrote that "the country for which you died wishes to forget the war in which you died. Its very name is a curse." Caputo goes on to observe that there were as yet no monuments, no statues in the town squares, no memorial plaques. These would serve as "reminders, and they would make it harder for your country to sink into the amnesia for which it longs."[9]

The reasons behind this amnesia in the French case have been examined, with extremely interesting and important results, by Professor Robert Frank and a group of young researchers at the University of Nanterre. Frank's hypotheses may be easily transposed to the United States.

Frank and his colleagues demonstrated that the French amnesia concerning the Algerian War is partial and not permanent and that the war is less forgotten than "taboo." Amnesia is after all, medically speaking, not incurable and is defined as a partial or total loss of memory, caused if not by injury or illness then by shock or psychological disturbance. Perhaps rather than forgetting (*oubli*), Frank contends, one should speak of a phenomenon of *occultation*, which might be translated as "cover-up."[10]

Polls and studies of the French press have shown that the French memory of events during the Algerian War is "selective," with arguably racist overtones. The nine ethnic Frenchmen and women killed at the Métro Charonne on February 8, 1962, are much more likely to be recalled than are the hundreds of Algerians (a conservative estimate) who were brutally murdered by the French police during and after the nonviolent demonstration on October 17, 1961. The memories that are most frequently covered up are precisely of the kinds of events that most distressed the intellectuals we have studied—the violence and torture, whether by the FLN, the OAS, or the French army and police.

The problem, according to Frank, is that the memory of the war of 1914–18 remains a uniting force among the French, whereas those of the wars of 1939–45 and 1954–62 are divisive, especially the latter. (The French war in Indochina has been almost completely forgotten.) In the case of the Second World War, despite the strange defeat of May–June 1940 and the tragedy and scandal of collaboration with the Nazi occupiers, there are heroes and glorious events that can be celebrated—Marc Bloch, Jean Moulin, de Gaulle himself, victories of the Free French such as that at Bir-Hakeim, the liberation of Paris, and the role of General Leclerc's Second Armored Division in it, the whole epic of the Resistance. Thus the Second World War has provided the French with prestigious names "with which we can baptize without shame our public squares, our avenues, our colleges, and our métro stations. But of the Algerian War what is left but the dead, easy to honor but almost impossible to commemorate."[11]

By "commemoration," Frank means a manifestation of collective memory that is not simple recollection, as in learning the facts for a test, for example. It is a public *rémemoration*, or remembrance, made by a group or groups, and is periodically expressed through a ceremony celebrating the events in question. And if one is going to have an effective commemoration of a war, several conditions must be met, none of which fit Algeria (or, needless to say, Vietnam). First, strong emotions must be generated and shared by a large majority of those present, so that the cermony can be made theatrical. This process can be easy when we are rendering honor to the dead.

But the public spectacle also must deliver a message to give a meaning to the emotions called forth, so that the ceremony can "resist the passage of time and transmit the sense of the event to generations who did not live it."[12]

At least in the first decades following a war, there must be present in the society a structured and organized group of veterans, like those of the Great War. The veterans can mobilize the authorities to be sure that the message is passed on. For a good commemoration, there must be harmony among the three principal components of collective memory, namely, the protagonists (while alive), the official memory (that is, the memory that the government in power wishes to convey), and the "public memory" of the society at large. If all of this works, we can have a kind of "commemorative chemistry," which has worked well for 1914–18 and reasonably well for 1939–45.

But it has not yet functioned for Algeria and probably cannot in the foreseeable future. Here there has been fascinating work done by Frédéric Rouyard on the bitter and ongoing debate among the French people, their elected representatives, and more specifically veterans and *pied-noir* groups, as to which date, if any, should be chosen to commemorate the end of the war. Should it be March 18 (the Évian accords), March 19 (the cease-fire ending an undeclared war), or a neutral date such as October 16, that of the return of the ashes of the first unknown soldier from Algeria, which in 1977 President Valéry Giscard d'Estaing preferred. Or alternatively, it could be moved to November 11, Armistice Day, which would become a kind of general "Memorial Day" (the French use the English phrase) to celebrate the ending of all of France's wars. As of this writing no decision has been made, and President François Mitterrand does not officially recognize any of the possible dates.[13]

Those who prefer the date of March 19, among them the FNACA, the leftist veterans' organization, are really celebrating a war that had no official existence and thus no real cause that could be associated with it. In a remarkable passage—which may be applied to the United States by changing only one word—Robert Frank presents with force and eloquence the dilemma faced by France after 1962:

> A war without a cause is a war without a message, and the *rémemoration* of a war without a message cannot be transformed into a true commemoration. The survivors can celebrate the fact that they did not die senselessly. But in honoring the memory of their dead comrades, they pose implicitly the frightful question, the most taboo by definition: Why did they die? The Algerian War only dragged on to make their comrades' sacrifice more vain. It is because this question is at bottom unbearable, impossible even to be asked, that this war cannot be commemorated.[14]

As Frédéric Rouyard has written, instead of leading to reconciliation, the commemoration of the Algerian War has remained a "theme of polemic." In this sense the Algerian War does not yet "belong to history."[15] Rouyard's words are echoed in George Herring's, cited in the Introduction to this book, that it may be many years before Vietnam moves "into the realm of history."

Can the French and we ourselves ever break out of this impasse? The terrible question of what to do with one's dead, both military and in the case of France also civilian, comes back again and again in the aftermath of the Algerian and Vietnam wars. It is a very difficult situation. For the *pieds-noirs* there are the ancestors who lie buried in what has become foreign soil and whose tombs must be maintained. Among the most important organizations formed since 1962 by the *pieds-noirs* now living in France are those whose principal goal is the care of the European cemeteries in Algeria.[16] Periodically some *pieds-noirs* return to what was once home, and they always go to the local cemetery to visit the family graves. One *pied-noir*, perhaps because the surroundings had changed and he had lost his bearings, could find only graves from his maternal side but not that of his paternal grandfather. Hence the memory of his presence in a land he had loved was effectively erased. "Where is the tomb of my grandfather, where is it? But where is my name in all of this? Where is it, in this historical error?" (*Mais où est-elle la tombe de mon grand-père? Mais où est mon nom là-dedans, où est-il dans cette erreur historique?*)[17]

Though no American soldiers are officially buried in Vietnam, the missing in action are part of our national obsession with the war, and the periodic return of the remains of American soldiers uncovered by the Vietnamese, which still continues fifteen years after the end of the war, is always noted in the press. Indeed, there is an ongoing debate as to whether the Vietnamese actually have in their possession more American remains, and release them in small numbers on carefully selected occasions to further their political goals.

Guy Pervillé, who has studied in depth Franco-Algerian relations since the granting of independence in 1962, accepts the "amnesty of individual responsibilities" for actions taken during the Algerian War as "necessary." On the governmental level General de Gaulle did indeed issue a two-stage amnesty. In 1966 there was a complete amnesty for antiwar activists, and those in exile were allowed to return to France. In 1968, probably as part of a bargain with the army to be assured of its loyalty during the May crisis of that year, de Gaulle pardoned the OAS activists who were in exile or still in prison,

including General Salan. (In the United States the partial
amnesty for draft resisters offered by President Gerald Ford
in September 1974 was more restrictive. It was followed by a
pardon for draft evaders, though not deserters from the
army, granted by President Jimmy Carter immediately after
his inauguration in January 1977.)

In the case of France and Algeria, Pervillé was looking
beyond the immediate political necessities that could be addressed by official governmental proclamations. He contends
that from this sort of just and appropriate amnesty there does
not inevitably follow "the amnesia of the collective responsibilities assumed or abdicated by the nation and its representatives." He believes that only a history of the Algerian War
that is rooted in a careful study of French colonization can
remove the passion from the Franco-française discord and
from the Franco-Algerian conflict by transcending them. Such
an effort is necessary in the two countries if future relations
between them are to be healthy and productive.[18]

Clearly, Pervillé's words apply to America and Vietnam
with only slight changes, and they suggest one way that in an
age of *dégagement*, intellectuals, at least scholars, can make a
public contribution. Through new perhaps as yet undevised
terminology and methodology, we need to find ways to make
these historical errors not necessarily palatable but under-
standable, fitting them into some kind of consensual context.
A different sort of engagement will probably be required,
given the immensity of the task, involving the kinds of group
work going on in France, where Algerian scholars are already
cooperating with their French colleagues. In the case of the
United States one would hope that eventually Vietnamese
intellectuals, both émigré and from reunited Vietnam, will
participate in these endeavors.

The semioticians speak of "engagement with the texts,"
and we can take a lesson from their book. As committed historians and citizens, we need to grapple with all varieties of
texts, the irreducible facts, the memoirs and the memories,

the newsreels, the entire vast array of historical materials, and help reconstruct these two divisive pasts. The goal of our engagement would be to bring these pasts into history, so that the mourning and the commemoration can proceed. Then at last we may be able to have amnesty without amnesia.

NOTES

Introduction

1. Bernard-Henri Lévy, *Éloge des Intellectuels* (Paris: Grasset, 1987), p. 48.

2. Quoted in Richard Bernstein, "Those 'New' Savants: Passé, or Past Their Prime?" *New York Times*, April 2, 1987, p. A4.

3. Lévy, *Éloge*, pp. 9–10, 12.

4. Ibid., p. 12.

5. David L. Schalk, *The Spectrum of Political Engagement: Mounier, Benda, Nizan, Brasillach, Sartre* (Princeton, N.J.: Princeton University Press, 1979), p. 116. See also pp. 24, 115.

6. The reviewer was Sol Gittleman, currently the provost of Tufts University. I am grateful to Professor Gittleman for sending me a copy of his text.

7. Though there are, of course, echoes of this debate in all the major biographies of Camus and Sartre, it has never been studied systematically, even in Germaine Brée's stimulating defense of Camus, written with verve and conviction: *Camus and Sartre: Crisis and Commitment* (New York: Dell, 1972). Some interesting background is provided in Claudie Broyelle and Jacques Broyelle, *Les illusions retrouvées: Sartre a toujours raison contre Camus* (Paris: Grasset, 1982).

8. *L'Éte fracassé*, 1973; *Couteau de Chaleur*, 1976; and *Fort Saganne*, 1980; all published in Paris by Éditions du Seuil. *Fort Saganne*, a powerful novel dealing with the penetration of the French army into the Saharan regions before World War I and drawing on the journals and letters of Gardel's grandfather, won the grand prix de l'Académie française, became a national best-seller, and was made into a movie starring Catherine Deneuve and Gérard Depardieu, regrettably not yet screened in this country.

9. Peter Novick, *That Noble Dream: The "Objectivity Question" and the American Historical Profession* (Cambridge: Cambridge University Press, 1988).

10. Carl N. Degler, "In Pursuit of American History," *American Historical Review* 92, no. 1 (February 1987): 3, 5. The text from Marc Bloch is his "Towards a Comparative History of European Societies,"

originally published in 1928 and reprinted in English in Frederic C. Lane and Jelle C. Riemersma, eds., *Enterprise and Secular Change: Readings in Economic History* (Homewood, Ill.: Irwin, 1953). Interestingly, Degler traces the origin of the notion of American exceptionalism to the famous question posed by the French essayist Crevècoeur, "Who is this new man, this American?"

11. Rémy Reiffel, "L'Empreinte de la guerre d'Algérie sur quelques figures intellectuelles 'de gauche'," *La Guerre d'Algérie et les intellectuels français*, Institut d'histoire du temps présent, cahier no. 10 (November 1988): 131–46. The quotation is from p. 131. Jeannine Verdès-Leroux, "La Guerre d'Algérie dans la trajectoire des intellectuels communistes," in ibid., pp. 211–23. Also see note 16.

12. See, for example, Sontag's interview with Charles Ruas, published in the *New York Times Book Review*, October 24, 1982, pp. 11, 39.

13. Peter Collier and David Horowitz, *Destructive Generation: Second Thoughts About the Sixties* (New York: Summit Books, 1989), p. 16. Collier and Horowitz also edited and published the proceedings of a symposium they organized, *Second Thoughts: Former Radicals Look Back at the Sixties* (Lanham, Md.: Madison Books, 1989). Although other issues besides Vietnam were cited, many of the participants mentioned that Vietnam was very important to their turning away from their previous beliefs.

14. Jean-François Sirinelli, "Une Histoire en chantier: L'Histoire des intellectuels," *Vingtième siècle*, no. 9 (January–March 1986): 103. The citation from de Tocqueville is taken from a letter to Beaumont dated December 1853.

15. Jacques Julliard, "Débat," in François Bédarida and Etienne Fouilloux, eds., *La Guerre d'Algérie et les chrétiens*, Institut d'histoire du temps présent, cahier no. 9 (October 1988): 140.

16. The IHTP organized a colloquium in Paris in April 1988 on this subject, and the results were published as cahier no. 10 in November 1988, cited in note 11. Other scholars working in the field include Anne Simonin on the editorial strategies of two antiwar publishing houses, Les Éditions de Minuit and les Éditions du Seuil; Jean-François Sirinelli on the "war of petitions" that raged during those years; Marie-Christine Granjon on Raymond Aron, Jean-Paul Sartre, and the Algerian conflict; and Etienne Fouilloux on Catholic intellectuals and the Algerian War.

17. Robert J. Glessing, *The Underground Press in America* (Bloomington: Indiana University Press, 1970). p. 10. Although many other topics were, of course, treated in this press, Glessing emphasized

that "moral resentment of the war in Vietnam" was a major factor in its rapid expansion in the 1960s.

18. John Newman, with Ann Hilfinger, eds., *Vietnam War Literature: An Annotated Bibliography of Imaginative Works About Americans Fighting in Vietnam*, 2nd. ed. (Metuchen, N.J.: Scarecrow Press, 1988).

19. John Gregory Dunne, "The War That Won't Go Away," *NYRB*, September 25, 1986, pp. 25–29. (A review of four books on the war)

20. Ibid., p. 29, citing a report in the *New York Times*.

21. *New York Times*, February 25, 1990, p. 43, "Campus Life" section.

22. Jonathan Mirsky, "The War That Will Not End," *NYRB*, August 16, 1990, pp. 29–36.

23. See, for example, Gordon O. Taylor, "American Personal Narrative of the War in Vietnam," *American Literature* 52, no. 2 (May 1980): 294–308.

24. Lance Morrow, "A Bloody Rite of Passage," *Time*, April 15, 1985, p. 26.

25. Ibid., p. 31.

26. Ibid. There is an extensive literature on the antiwar movement considered in its mass aspects. See especially Charles De-Benedetti, *An American Ordeal: The Antiwar Movement of the Vietnam Era* (Syracuse, N.Y.: Syracuse University Press, 1990). DeBenedetti tends to answer Lance Morrow's question in the affirmative.

27. Karen J. Winkler, "The Vietnam War Scores Well at the Box Office, but It Fails to Attract Many Researchers," *Chronicle of Higher Education*, September 30, 1987, pp. A4–A5.

28. George C. Herring, "America and Vietnam: The Debate Continues," *American Historical Review* 92, no. 2 (April 1987): 350.

29. Ibid., p. 361.

30. Ibid.

31. Philip Caputo, *A Rumor of War* (New York: Ballantine, 1978), p. xx.

32. Daniel Berrigan and Robert Coles, "A Dialogue Underground," *NYRB*, March 11, 1971, p. 24.

33. Martha Ritter, "Echoes from the Age of Relevance," *Harvard Magazine*, July–August 1981, p. 10.

Chapter 1

1. See Peter Dale Scott, "Tonkin Bay: Was There a Conspiracy?" *NYRB*, combined issue, January 29, 1970, pp. 31–41. This is

an extensive and careful evaluation of the evidence, which concludes that American deception was involved. See also Joseph C. Goulden, *Truth Is the First Casualty: The Gulf of Tonkin Affair—Illusion and Reality* (Skokie, Ill.: Rand McNally, 1970).

2. John Talbott, *The War Without a Name: France in Algeria, 1954–1962* (New York: Knopf, 1980), p. 61.

3. Bernard Droz and Evelyne Lever, *Histoire de la guerre d'Algérie* (Paris: Éditions du Seuil, 1982), p. 291. Even within the paratroopers (always known affectionately or fearfully, depending on one's politics, as the *paras*), there was a hierarchy based on the color of the beret. Green berets were worn by the foreign legionaries, and red by the colonial paras. These were much more highly regarded than the "inglorious 'blue berets,' the paras from metropolitan France." See also Bernard B. Fall, "Vietnam: The Undiscovered Country," *NYRB*, March 17, 1966, p. 8.

4. The literature on pacification in both wars is vast. An exceptionally perceptive and illuminating account by nine soldiers, three of whom were officers, was published under the title "La Pacification" in *Esprit* 29, no. 291 (January 1961): 7–24. These writers believed that the military command was deceiving itself, that pacification was a fallacy, that only the walls were pacified. The French military grip might be tighter by 1960, and the large ALN units broken up, but politically there had been no progress. To pacify a region meant to send troops. "It remains 'pacified' as long as the troops are present" (p. 11).

5. D. A. N. Jones, "The Monstrous Thing," *NYRB*, December 17, 1964, p. 8. Before Leuillette's memoir was published in book form, excerpts appeared in *Esprit* 27, no. 272 (April 1959): 562–68. It is noteworthy, I believe, that the editors of *Esprit* were the first to recognize the significance of this document. See Chapter 3.

6. Jacques Massu, *La Vraie Bataille d'Alger* (Paris: Plon, 1971). General Jacques Paris de Bollardière, who would not authorize torture in the district in Algeria under his control and whose case caused a scandal in France, finally broke his silence and published a moving memoir as a direct response to his old comrade Massu: *Bataille d'Alger: Bataille de l'homme* (Paris: Declée de Brouwer, 1972). General Paris de Bollardière was imprisoned in 1957 after refusing an order to search the mosques in his district and, in a painful break with a long family tradition, eventually left the army in 1961. Jules Roy, a former air force colonel and a *pied-noir* who had resigned from active duty because of the torture already practiced by the French forces

in Indochina before 1954, wrote a searing deunciation entitled *J'Accuse le Général Massu* (Paris: Les Éditions du Seuil, 1972).

7. D. A. N. Jones, "The Monstrous Thing," p. 9.

8. Ibid.

9. George C. Herring, *America's Longest War: The United States and Vietnam, 1950–1975* (New York: Wiley, 1979), p. 142.

10. Philippe Ivernel, *"Paris-Match* à l'heure du cessez-le-feu," *Esprit* 30, no. 307 (June 1962): 981.

11. Anthony Lake, "The End of an Analogy," *Boston Review* 9, no. 1 (February 1984): p. 13. (Italics his)

12. Sartre's role as perhaps the leading '*tiersmondiste*' intellectual will be described in Chapter 3. Sartre came to this conviction earlier than many, at least by 1958, writing that from the first colonial rebellions at the end of World War II intellectuals should have realized that they were observing the beginning of "what would become the most considerable event of the second half of this century: the awakening of nationalism among the Afro-Asian peoples": "Les Grenouilles qui demandent un roi," first published in *L'Express,* September 25, 1958, and reprinted in *Situations, V: Colonialisme et néo-colonialisme* (Paris: Gallimard, 1964). The citation is from p. 121. See also Bernard Droz's discussion of the Algerian War as the only example in French history of an "episode of decolonization that took on the depth of a national drama," in "Le Cas très singulier de la guerre d'Algérie," *Vingtième siècle*, no. 5 (January–March 1985): 81.

13. Paul Mus, *Guerre sans visage: Lettres commentées du sous-lieutenant Émile Mus* (Paris: Éditions du Seuil, 1961), p. 168. Robert Aron, François Lavagne, Janine Feller, Yvette Garnier-Rizet eds., *Les Origines de la guerre d'Algérie* (Paris: Fayard, 1962). In his introduction to this collection of texts and documents, Aron presents a conservative view that when in the future an attempt is made to write objectively the history of the colonial era, Algeria will be judged as one of the "finest successes of this necessary phase of world evolution." Sounding more like a *tiersmondiste* than he doubtless would have liked to admit, Aron concludes that there was in colonization an "internal contradiction that could not but end with its destruction." The end of Algérie française was not primarily due to the struggles of the Europeans or the Muslims of Algeria but resulted from "a historical process that the entire world has undergone" (pp. 7–8).

14. Cf. Garry Wills's observation that the reality of Vietnam was never grasped by Americans, who viewed it more as "a social ther-

mometer for this country." In Wills's view the Vietnam War was not particularly American but, rather, a "small (but unnaturally inflated) episode" in the larger historical process of decolonization: "The Meaning of Vietnam," *NYRB*, June 12, 1975, p. 24.

15. Stanley Hoffmann, "Vietnam: An Algerian Solution?" *Foreign Policy*, no. 2 (Spring 1971): 4.

16. Similarities in the economic impact of the two wars, including the escalating patterns of military spending that had inflationary effects in both cases, were explored in Jean-Charles Asselain, "'Boulet colonial' et redressement économique (1958–1962)," in Jean-Pierre Rioux, ed., *La Guerre d'Algérie et les Français* (Paris: Fayard, 1990), pp. 289, 652.

17. Raymond Aron, *L'Algérie et la république* (Paris: Plon, 1958), p. 113.

18. Christopher Lasch, "New Curriculum for Teach-Ins," *The Nation*, October 18, 1965, p. 240. Eisenhower was, of course, ineligible and in retirement, and so the task of settling Vietnam was left to the man who had served as his vice-president from 1953 to 1961.

19. Ronald Steel, "A Visit to Washington," *NYRB*, October 6, 1966, pp. 5–6.

20. John Gerassi, "Trouble at San Francisco State," *NYRB*, April 11, 1968, p. 45.

21. Michel Winock, *La République se meurt: Chronique 1956–1958* (Paris: Éditions du Seuil, 1978), p. 11. See "T.M." (editors of *Les Temps modernes*), "He Disgraces the Name of Socialism," *Les Temps modernes* 12, no. 136 (June 1957): 1884.

22. See Sandy Vogelgesang, *The Long Dark Night of the Soul: The American Intellectual Left and the Vietnam War* (New York: Harper & Row, 1974), pp. 1, 114–15, 202.

23. Talbott, *The War Without a Name*, p. 58. Talbott goes on point out the irony inherent in the fact that the Socialist party, "the party of Jean Jaurés, Léon Blum, and other enemies of colonialism and war, controlled the government conducting the biggest and most expensive colonial war in France's history. The war divided the Socialists as the war in Indochina later divided Lyndon Johnson's Democrats in the United States" (p. 74).

24. This speech has been analyzed and cited in several places, for obvious reasons, because even though it was not especially shocking when it was delivered, it became politically embarrassing for France's future president. The first part of the citation is correct, but the more damning phrase, "The only negotiation is war," does not appear in any of the reports of the speech. See Catherine Nay,

Le Noir et le rouge (Paris: Grasset, 1984), p. 215. The speech was reprinted in part in *Le Monde*, November 14–15, 1954, p. 3; in *L'Année politique 1954* (Paris: Presses Universitaires de France, 1955), pp. 276–77; and in Aron et al., *Les origines de la guerre d'Algérie*, p. 321.

25. Cf. Droz and Lever, *Histoire de la guerre d'Algérie*, p. 155.

26. See Pierre-Henri Simon, *Contre la torture* (Paris: Éditions du Seuil, 1957), p. 77; Henri Alleg, *La Question* (Paris: Éditions de Minuit, 1958), p. 36; and Droz and Lever, *Histoire de la guerre d' Algérie*, p. 299.

27. Andrew Kopkind, "The Thaw," *NYRB*, April 25, 1968, p. 5. Kopkind was not alone in drawing the Robert Kennedy–Charles de Gaulle parallel. See Paul Goodman, "We Won't Go," *NYRB*, May 18, 1967, p. 17. The Vietnam protest movement hoped for a fundamental reconstruction of American society: "It is not enough to get out of Algeria or Vietnam and end up with De Gaulle—or Bobby Kennedy."

28. Henry Steele Commager, "The Case for Amnesty," *NYRB*, April 6, 1972, p. 23.

29. Jean Lacouture, "Les camelots du Président Johnson," *Le Nouvel Observateur*, January 5–11, 1966, pp. 2, 4.

30. This remarkable example of the soporific or sedative use of history is from Richard Neustadt, "Uses of History in Public Policy," *Humanities* 2, no. 5 (October 1981): 1.

31. Letter from Professor Neustadt to the author, May 14, 1982.

32. Henry Kissinger, *The White House Years* (Boston: Little, Brown, 1979), p. 228. Kissinger's choice of the verb *extricate* is most interesting.

33. Simon Head, "Story Without End," *NYRB*, August 9, 1973, p. 26.

34. Daniel Ellsberg claimed in 1971 that three years earlier Kissinger frequently said "in private talks that the appropriate goal of U.S. policy was a 'decent interval'—two to three years—between the withdrawal of U.S. troops and a communist takeover in Vietnam": "Laos: What Nixon Is Up To," *NYRB*, March 11, 1971, p. 13. See also Hoffmann, "Vietnam: An Algerian Solution?" p. 30; and I. F. Stone, "A Bad Deal That May Not Work," *NYRB*, November 30, 1972, p. 7. With his usual acuity Stone points out the possibility, even before the pending cease-fire agreements were signed, that the accords would be "legalistic eyewash . . . to keep Thieu happy while Nixon brings the POW's home and after a 'decent interval' stands by and lets the other side take over." For a later prediction, after the January 1973 cease-fire, see William Shawcross, "How Thieu

Hangs On," *NYRB*, July 18, 1974, pp. 16–19. Shawcross, who had just returned from a visit to South Vietnam, believed (correctly) that the GVN would be lucky if it survived another year.

35. Alastair Horne, *A Savage War of Peace: Algeria 1954–1962* (New York: Penguin Books, 1979), p. 396.

36. Ibid., p. 475.

37. Ibid., p. 507.

38. Ibid., p. 548.

39. The young draftee Pierre Boisgontier, whose father, an officer, had been killed in the Second World War, offered to do civilian alternative service instead of comfortably remaining in France proper in the army, as was his legal right. He was jailed instead. See Jacques Tinel, "Construire des choses solides et vrais," *Esprit* 28, no. 260 (December 1960): 2061–64. In the American case, if a Selective Service registrant's father had been killed in World War II or Korea, of if his brother or father were in Vietnam, he was exempted from the draft. See also "T.M." (editors of *Les Temps modernes*), "Qui démoralise l'armée?" *Les Temps modernes* 11; no. 123 (March–April 1956): 1535–36.

40. See the article on Ky in *Current Biography Yearbook 1966* (New York: H. W. Wilson, 1967), p. 230. Also Jean Lacouture, "Vietnam: The Turning Point," *NYRB*, May 12, 1966, p. 7. In his autobiographical memoir, Ky himself stated that after his air force training in France in 1954 he went to Algeria for "five months bombing and strafing training." He does not mention any actual combat duty. See Ky's *How We Lost the Vietnam War* (New York: Scarborough Books, 1978), p. 18.

41. Robert Lacoste, interview with Jean Daniel, April 22, 1958, reprinted in Jean Daniel, *De Gaulle et l'Algérie* (Paris: Éditions du Seuil, 1986), pp. 42–43. (Originally published in *L'Express*, May 2, 1958)

42. Horne, *A Savage War of Peace*, p. 321. Cf. also, Herring, *America's Longest War*, pp. 153–54.

43. Philip Caputo, *A Rumor of War* (New York: Ballantine, 1978), pp. xix, 160, 294.

44. Albert-Paul Lentin, "L'Algérie au jour le jour," *Les Temps modernes* 14, nos. 156–57 (February–March 1959): 1245.

45. Horne, *A Savage War of Peace*, p. 546. Cf. Bernard B. Fall, *Last Reflections on a War* (New York: Doubleday, 1967), p. 221. This is a brilliant posthumous collection edited by his widow after Fall was killed while covering the Vietnam War. He notes that the French in Algeria had a comfortable, 11–to–1 troop ratio. And (unlike the

Americans in Vietnam, who despite massive efforts were never able to prevent troops and weapons from entering South Vietnam) they effectively sealed off the border between Algeria and Tunisia. They were, however, winning at the price of being the "second-most-hated country in the world, after South Africa, in the United Nations." They were tying down most of the contingent of draftees and thus not making their contribution to the defense of continental Europe. They were spending $3 million a day for eight years. "The price also included two mutinies of the French Army and one overthrow of the civilian government. At that price the French were winning the war in Algeria, *militarily*. The fact was that the military victory was totally meaningless. This is where the word grandeur applies to President de Gaulle: He was capable of seeing through the trees of military victory to a forest of political defeat and he chose to settle the Algerian insurgency by other means." (Italics his)

46. The former phrase was used by Philip Caputo to describe some of his fellow officers, the latter by Bernard Fall.

47. Etienne Fouilloux, "Ordre social chrétien et Algérie française," in *La Guerre d'Algérie et les chrétiens*, Institut d'histoire du temps présent, cahier No. 9 (October 1988): 83. See also the section "Intellectuals," in Mary McCarthy, *The Seventeenth Degree* (New York: Harcourt Brace Jovanovich, 1974), pp. 122–30.

48. Fall, "Vietnam, the Undiscovered Country," pp. 8–9. Fall goes on to note that the centurion complex can ultimately lead the military elite to take military action at home, as it did in France. So far in the American case it was limited to soldiers' threatening to beat up demonstrating college students when they finished their tour of duty in Vietnam. But "like the French paratroopers in Algeria," Americans soldiers had the potential of evolving to a point at which they would move to destroy what they had "supposedly been sent to defend: Freedom of political choice."

49. Lentin, "L'Algérie au jour le jour," p. 1236.

50. For a classic portrayal of this terrible problem in Vietnam, which echoes almost verbatim the soldiers' accounts published in *Esprit* and *Les Temps modernes* during the Algerian War, see Daniel Lang, *Casualties of War* (New York: McGraw-Hill, 1969), esp. pp. 19–20. Pfc. Sven Eriksson (a pseudonym) told Lang, "From one day to the next, you could see for yourself changes coming over guys on our side—decent fellows, who wouldn't dream of calling an Oriental a 'gook' or a 'slopehead' at home. But they were halfway around the world now, in a strange country, where they couldn't tell who was their friend and who wasn't."

51. The literature here from both wars is voluminous, and some of it will be cited in later chapters. French antiwar periodicals published many reports by draftees from the even larger number submitted to them. They all corroborate one another, and long before the admissions of Lieutenant Leuillette and General Massu pitilessly document the use of torture, reprisals, the summary execution of nomads caught in a *zone interdite*, simple unprovoked murder, the destruction of villages, the killing of animals, and the burning of crops to deny sustenance to the FLN, while in the process bringing the local population to even greater misery. Two of the most powerful of these examples are an account by the anonymous soldier "X," "Journal de campagne," *Les Temps modernes* 13, nos. 137–38 (July–August 1957): 160–73; and Robert Bonnaud, "La paix des Nementchas," *Esprit* 25, no. 249 (April 1957): 580–92. This article was so explosive and so controversial in its charges that the editors of *Esprit* inserted a special note indicating that "Robert Bonnaud" was not a pseudonym but a real veteran who had just returned from duty in Algeria and that they stood by his testimony. For the comparison with Vietnam, see Telford Taylor, *Nuremberg and Vietnam: An American Tragedy* (New York: Bantam, 1971), p. 145.

52. Cf. Daniel, *De Gaulle et l'Algérie*, p. 145, from an article originally published in March 1960. An officer proudly told General de Gaulle that he had "completely cleaned up" his sector. When de Gaulle asked him, "And if we pull out the troops?" the officer replied. "Then everything would begin again."

53. Cf. François Sarrazin, "L 'Algérie, pays sans loi," *Esprit* 23, nos. 230–31 (September–October 1955): 1626; Droz and Lever, *Histoire de la guerre d'Algérie*, pp. 139, 288; Georges Houdin, "Ce que fut l'attitude des chrétiens français," *La Nef*, cahiers nos. 12–13 (October 1962–January 1963): 77–78; and Frances FitzGerald, *Fire in the Lake: The Vietnamese and the Americans in Vietnam* (Boston: Little, Brown, 1972), p. 345. See also Taylor, *Nuremberg and Vietnam*, pp. 146–47, 152, 171. In 1967 a U.S. official in Saigon, who would have had every reason to keep the number as low as possible, estimated in a press briefing that 1.5 million refugees had been created by the war: McCarthy, *The Seventeenth Degree*, p. 94.

54. Jean-Philippe Talbo-Bernigaud, "Zones interdites," *Les Temps modernes* 16, no. 177 (January 1961): 708–26. Talbo-Bernigaud observes that in 1959 the French military authorities in Algeria decreed a name change for these areas, calling them *champs de tir de circonstance*, which theoretically means that one would be more careful about whom one was shooting. However,

the soldiers still spoke of *zones interdites*, and it was easy to get authorization to fly in and strafe any group of people discovered in
these zones. On "free-fire" or "free-strike" zones, see also Taylor,
Nuremberg and Vietnam, pp. 145–46; and Herring, *America's Longest
War*, p. 153.

  55. In Cecil Woolf and John Bagguley, eds., *Authors Take Sides
on Vietnam* (New York: Simon & Schuster, 1967), p. 30. The French
had a "diplomatic translation" for napalm, calling the containers
bidons spéciaux, which might be rendered in English as "special can-
isters": Talbo-Bernigaud, "Zones interdites," p. 718.

  56. Cf. FitzGerald, *Fire in the Lake*, pp. 385, 423. Also Gloria
Emerson, *Winners and Losers. Battles, Retreats, Gains, Losses and
Ruins from the Vietnam War* (New York: Harcourt Brace Jovanovich, 1976), pp. 240–41. And especially the extremely detailed and
carefully researched study by Alfred W. McCoy, Cathleen B. Read,
and Leonard P. Adams II, *The Politics of Heroin in Southeast Asia*
(New York: Harper & Row), 1972. See also Alfred W. McCoy, "A
Correspondence with the CIA," *NYRB*, September 21, 1972,
pp. 26–35. McCoy reprints letters pertaining to the CIA's efforts
to stop the publication of his book. One of the statistics that
McCoy cites here is that in 1969, before a significant number of
GIs started using heroin in Vietnam, there were an estimated
300,000 addicts in the United States. Early in 1972 the government estimate was 600,000 (p. 35).

  57. Bernard W. Sigg, *Le silence et la honte: Nevroses de la guerre
d'Algérie* (Paris: Messidor, 1989), pp. 50–51. This is an important and
pioneering work, the first such study to appear in France. Sigg is a
psychoanalyst who served in Algeria as a medic. He writes that alcoholism was one of the great scourges of the Algerian War, both
during and after the conflict; he treated many cases of delirium
tremens. There is, however, no available research in France on the
degree to which other drugs were utilized and on their impact. Sigg
regrets that there are no parallels in the medical and sociological
literature in France with "the mass of documentation and studies
prepared by academics and administrators in the United States dealing with drug addiction among the GI's in Vietnam." See also Roy,
J'Accuse le Général Massu, p. 44.

  58. Cf. Emerson, *Winners and Losers*, pp. 25, 385. The American
euphemism was "The Bell Telephone Hour." When de Gaulle came
to power in 1958, he tried with only minimal success to halt such
methods, for which he personally had nothing but contempt. He
told officers serving in Algeria to use telephones to speak into rather

than to make people speak. For an exceptionally gripping account by a common soldier who served in Algeria in 1956 and 1957, see Jacques Pucheu, "Un an dans les Aurès," *Les Temps Modernes* 13, no. 123 (September 1957): 433–47. For a comparison of the use of torture in Vietnam and in Algeria, see Bernard B. Fall, *Viet-Nam Witness: 1953–1966* (New York: Praeger, 1966), pp. 300–3.

59. Roy, *J'Accuse le Général Massu*, p. 43 Roy argues that Massu may have won the battle of Algiers, but "you contributed to the loss of the Algerian War because your analysis of the situation was just as false as that which the Pentagon is making today regarding Vietnam."

60. The chef was André Daguin and the town was Auch where Daguin now has a two-star restaurant. Cf. Robert Daley, "Jet-Set Chefs," *New York Times Magazine*, February 6, 1977, p. 76. The American example is taken from Emerson, *Winners and Losers*, p. 322.

61. For a very thoughtful analysis of the ideology behind the Algerian revolution, which he categorizes as "Arabo-Islamism," see Daniel, *De Gaulle et l'Algérie*, p. 156. (From an article originally published in May 1960) Daniel, himself a *pied-noir*, who had close contacts with all sides in the struggle, came to the conclusion that Algerian independence was both necessary and just.

62. All of the histories in French and English devote attention to this painful return, and an extensive literature describing and evaluating the *pied-noir* implantation in France has sprung up since 1962. A good account in English of this "exodus" may be found in Horne, *A Savage War of Peace*, pp. 505–34.

63. Ali Haroun, *La 7e Wilaya. La Guerre du FLN en France, 1954–1962* (Paris: Éditions du Seuil, 1986), p. 307. *Wilaya* is the Arabic word for region, and the FLN and its military arm, the ALN, divided Algeria into six *wilayas*, with the seventh nominally being France itself. Haroun, who was one of the five leaders of the FLN in France during the war and was never caught, had access to important documents. He provides careful tabulations of the very significant sums that were successfully smuggled out of France for the FLN war chest.

64. Ibid., p. 327.

65. Caputo, *A Rumor of War*, p. 26.

66. George C. Herring, " 'Peoples Quite Apart': Americans, South Vietnamese, and the War in Vietnam," *Diplomatic History* 14, no. 1 (Winter 1990): 8.

67. Ibid.

68. Talbott, *The War Without a Name*, p. 94.

69. Jean-Marie Domenach, "Culpabilité collective," *Esprit* 25, no. 254 (October 1957): 403.

70. In Woolf and Bagguley, eds., *Authors Take Sides on Vietnam*, p. 69. (Italics hers)

Chapter 2

1. In David L. Schalk, *The Spectrum of Political Engagement: Mounier, Benda, Nizan, Brasillach, Sartre* (Princeton, N.J.: Princeton University Press, 1979), esp. pp. 5–6, 110–16, I review some of the efforts that intellectuals have made to define themselves. Among the most important works that discuss this subject are Victor Brombert, *The Intellectual Hero* (Philadelphia: Lippincott, 1961); Philip Rieff, ed., *On Intellectuals* (New York: Doubleday, 1970); Louis Bodin, *Les Intellectuels* (Paris: Presses Universitaires de France, 1964); Pascal Ory and Jean-François Sirinelli, *Les Intellectuels en France, de l'affaire Dreyfus à nos jours* (Paris: Armand Colin, 1986).

2. Sandy Vogelgesang, *The Long Dark Night of the Soul: The American Intellectual Left and the Vietnam War* (New York: Harper & Row, 1974), p. 14.

3. Gramsci's well-known distinction between "traditional" and "organic" itellectuals is thoughtfully discussed in James Joll, *Antonio Gramsci* (New York: Penguin Books, 1977), pp. 120–24. See also Gramsci's *Selections from the Prison Notebooks*, trans. Quintin Hoare and G. N. Smith (New York: International Publishers, 1971), pp. 5–6, 9, 15–16.

4. I find this terminology more useful than Gramsci's conception of the "organic" intellectual, the latter being in my view too closely linked to a single dominant class. The class structure in advanced Western societies has probably become more fuzzy than it was when Gramsci languished in a Mussolini prison and composed his brilliant notebooks.

5. Edward Shils, "The Intellectuals and the Powers," in Rieff, ed., *On Intellectuals*, p. 33.

6. Ibid., p. 46.

7. See especially William F. Johnston, "The Origin of the Term 'Intellectuals' in French Novels and Essays of the 1890's," *Journal of European Studies* 4 (1974): 43–56.

8. Jean-Marie Colombani, "Un poujadisme démocratique," *Le Monde hebdomadaire*, June 7–13, 1990, p. 8.

9. *Time*, November 15, 1971, p. 40; William Pfaff, "L'Homme engagé," *The New Yorker*, July 9, 1990, pp. 83–91.

10. In the *Poughkeepsie Journal*, January 9, 1987, p. 4A and nationally syndicated.

11. Richard Bernstein, "Revisionists and Storytellers—Is It Passé to Be Engagé?" *New York Times Book Review*, January 5, 1986, pp. 3, 29.

12. Erica Abeel, "Hers," *New York Times*, January 25, 1979, p. C2.

13. Hervé Hamon and Patrick Rotman, *Les Porteurs de valises: La Résistance française à la guerre d'Algérie* (Paris: Albin Michel, 1979).

14. Paul Clay Sorum, *Intellectuals and Decolonization in France* (Chapel Hill: University of North Carolina Press, 1977), p. xiii.

15. In speaking of student activism, Sorum does make the claim, which later research by French scholars partially questions, that in contrast with "the protests of American students against the American war in Vietnam, the lack of student activity against the Algerian war is striking" (ibid., p. 177).

16. The other exception was my own contribution, which was later published under the rubric of an "intervention" rather than a formal paper. French colleagues found the tentative parallels I drew extremely interesting and several, using almost identical language, told me that French inattention to such questions derived from "our limited vision and tendency to focus only on the affairs of the hexagon [i.e., metropolitan France]."

17. Rémy Reiffel, "L'Empreinte de la guerre d'Algérie sur quelques figures intellectuelles 'de gauche'," *La Guerre d'Algérie et les intellectuels français*, Institut d'histoire du temps présent, cahier no. 10 (November 1988): 142.

There is one other mention of Vietnam in Reiffel's fascinating article, but it is in the context of French intellectual opposition to the American war in Vietnam, which will not be studied here. When Reiffel asked his respondents about their later interventions, they seemed perplexed and not sure how to reply, though the mathematician Laurent Schwartz did see in his "combat against the war in Vietnam a resurgence of that earlier form of struggle" (ibid., p. 145).

18. Vogelgesang, *The Long Dark Night of the Soul*, p. 64.

19. These statistics are from Paul Lauter and Florence Howe, "The Draft and Its Opposition," *NYRB*, June 20, 1968, p. 29. On the Resist movement, see *NYRB*, January 28, 1971, p. 11.

20. Vogelgesang, *The Long Dark Night of the Soul*, pp. 35, 130.

21. Cited in Jessica Mitford, *The Trial of Dr. Spock* (New York: Vintage Books, 1970), p. 49. Raskin was a member of the "Boston Five," defendants in the Spock trial of 1968, one of the most famous political trials of the Vietnam era.

22. Francine du Plessix Gray, *Divine Disobedience: Profiles in Cath-*

olic Radicalism (New York: Vintage Books, 1971), p. 52; Francine du Plessix Gray, "Address to the Democratic Town Committee of Newtown, Conn.," *NYRB*, May 6, 1971, p. 18. See also Anne Klejment, *The Berrigans* (New York: Garland Press, 1979), p. xx.

23. Daniel Berrigan, *Consequences, Truth, and . . .* (New York: Macmillan, 1971), p. 55.

24. Paul Goodman, "Appeal," *NYRB*, April 6, 1967, p. 38.

25. The standard work on these riots, which draws on extensive interviews with veterans, is Jean-Pierre Vittori, *Nous, les appélés d'Algérie* (Paris: Stock, 1977). For a bitter attack on the French Communist party for its role in subduing and channeling this agitation, in preventing its spread and hence aiding in the full French mobilization, and in preparing the "battle of Algiers" that began the next year, see Edgar Morin, *Autocritique* (Paris: Éditions du Seuil, 1975), pp. 186–90.

26. Paul Goodman, "We Won't Go," *NYRB*, May 18, 1967, p. 17.

27. Paul Goodman, "A Causerie at the Military–Industrial," *NYRB*, November 23, 1967, p. 18.

28. See Jean-François Sirinelli, "Guerre d'Algérie, guerre des pétitions?" in Rioux, *La Guerre d'Algérie et les intellectuels français*, pp. 181–210.

29. As a historian educated in America in the late 1950s and early 1960s my training and thinking have always been linear and progressive, and I have had difficulty accepting what I was discovering. But the evidence, as will be seen, is convincing.

30. In the French case the reasons lie principally with the lack of full recovery from the Second World War, with the fact that the "thirty glorious years" of the French economic miracle were only beginning, and with the localized ideological struggle generated by the cold war and the Marshall Plan. France's own war in Vietnam, which was finally settled in July 1954, had much less impact on its intellectual class than on issues in conflict lying within the boundaries of Europe.

With regard to the United States the extreme *dégagement* of the Eisenhower years was partially due to the tremendous impact of Daniel Bell's *The End of Ideology* (1960), preceded by that of David Riesman's *The Lonely Crowd* (1950). The budding intelligentsia of my generation at Harvard (1958–63) thought that we fully understood Riesman's terminology and lamented it, as it seemed to indicate that we (or everyone else but us) were so "other directed" that we were passively letting ourselves be controlled by the forces of contemporary technological society.

31. On this committee, see the account by Edgar Morin, one of

its founders, in his *Autocritique*, p. 187; also the unsigned articles reporting on the establishment and activities of the committee in *Le Monde*, November 6–7, 1955, p. 4; and November 9, 1955, p. 4.

32. See Eric R. Wolf and Joseph G. Jorgensen, "Anthropology on the Warpath in Thailand," *NYRB*, November 19, 1970, p. 35.

33. Sirinelli, "Guerre d'Algérie, guerre des pétitions?" p. 189.

34. Here perhaps the right side of the brain has won the day, in that the virulence of the charges of national betrayal or treason made by conservative intellectuals, especially though not solely in France, has obscured the patriotic ingredient often present at all levels of antiwar engagement, even the most militant.

35. Pierre-Henri Simon, "Opération bonne conscience," *Esprit* 25, no. 253 (September 1957): 245.

36. Mary McCarthy, *The Seventeenth Degree* (New York: Harcourt Brace Jovanovich; 1974), p. 317.

37. Norman Mailer, *The Armies of the Night: History as a Novel: The Novel as History* (New York: New American Library, 1968), p. 129.

38. Gray, "Address to the Democratic Town Committee of Newtown, Conn.," p. 19. This statement happens to have been made by an American writer of French Catholic extraction, in discussing the antiwar engagement of the Berrigan brothers, Daniel and Philip. As will be seen in Chapter 3, almost identical statements were made a decade earlier by French Catholic opponents of the Algerian War. Gray further defines civil disobedience as "based on the idea that you break small laws—like defying lunch counter segregation laws, or bus segregation laws, or antistrike laws, or laws about draft files, or nineteenth-century laws forbidding you to shelter runaway slaves—to point out the existence of higher laws, like the brotherhood of man or the atrocity of war."

39. See Schalk, *The Spectrum of Political Engagement*, esp. pp. 22–24.

40. Hans J. Morgenthau, "Reflections on the End of the Republic," *NYRB*, September 24, 1970, p. 38. For a good discussion of this controversial debate—to this day there is uncertainty as to who was the victor—see Melvin Small, *Johnson, Nixon, and the Doves* (New Brunswick, N.J.: Rutgers University Press, 1988), pp. 50–51.

41. When Morgenthau reread the essays he had written for the *NYRB* and other journals in the 1960s, he was struck by their activist and rationalist tone. "One only needed, or so it seemed, to call the President's attention to the probable consequences of certain policies and show him the alternatives and their probable consequences, and

he would choose a policy most likely to serve the national interest":
"Reflections on the End of the Republic," p. 38.

42. Hans J. Morgenthau, "Vietnam: Shadow and Substance,"
NYRB, September 16, 1965, p. 3. Cf. Howard Zinn, *Vietnam: The
Logic of Withdrawal* (Boston: Beacon Press, 1967), p. 108. In developing his argument for the American withdrawal from Vietnam,
Zinn cites historical examples of when nations have pulled out of
losing ventures and not lost prestige. "More recently, and more pertinently, France moved out voluntarily [*sic*] from Algeria and from
Indochina; today she has more prestige than ever before."

43. Morgenthau, "Vietnam, Shadow and Substance," p. 5.
Morgenthau's comment struck a sensitive nerve, as I ponder the
suffering of American Vietnam veterans, their extremely high suicide rate, the epidemic of drug addiction they brought home, the
high percentage of America's homeless whom they represent, the
ongoing Agent Orange tragedy, and so many more evil residues
in our society that can be directly traced to our Vietnam war. I
doubt whether I am alone in these kinds of painful, even anguished reflections. The very day I completed this note I took a
short trip to New York City from Poughkeepsie. On the train an
obviously emotionally distraught black veteran was shouting about
how he would never go back to Vietnam, would never kill Asians
again or fight to defend white people, and would take up weapons only to defend black Africans.

44. For many American intellectuals—at least those who are not
natives of Pennsylvania—Altoona as a middle-sized industrial city
symbolizes the hawkish American spirit that, if it did not lead us into
the Vietnam disaster, arguably helped keep us there until 1975. One
thinks of the magnificent and gripping Michael Cimino film, *The
Deer Hunter* (1978), much of which takes place in an industrial city
of Pennsylvania and which answers Norman Mailer's question "Why
are we in Vietnam?" better than any single work of art I know. *The
Deer Hunter*, like any important work of art, is subject to conflicting
interpretations. In my view it is fundamentally an antiwar document,
patriotic perhaps but never militaristic or chauvinistic.

45. Elizabeth Hardwick, "We Are All Murderers," *NYRB*, March
3, 1966, pp. 6–7.

46. "Le plus difficile," *Esprit* 27, no. 278 (November 1959): 470.

47. In reporting on street battles in Rome in November 1960,
between groups of students shouting "Algérie française," and "Algérie algérienne," Jean-Marie Domenach observed, "We Frenchmen
have the gift of universalizing our political dissensions and of ex-

porting our problems of conscience": "La part de l'avenir," *Esprit* 29, no. 291 (January 1961): 144.

48. Quoted in the preface to Alistair Horne, *A Savage War of Peace: Algeria 1954–1962* (New York: Penguin Books, 1979), p. 13.

49. George Steiner, Letter to Noam Chomsky, published in the *NYRB* March 23, 1967, p. 28. Steiner goes on to note that he is not trying to score a debater's point with Chomsky but that he writes "in deep personal perplexity."

50. Hamon and Rotman, *Les Porteurs de valises*, p. 82.

51. One of the principal streets in Algiers is now named in his honor.

52. Francis Jeanson, "Para-Pacification," *Esprit* 25, no. 250 (May 1957): 817.

53. Noam Chomsky, Letter to George Steiner, published in the *NYRB* March 23, 1967, p. 28.

54. Ibid., p. 28.

55. Noam Chomsky, "After Pinkville," *NYRB*, January 1, 1970, p. 14.

56. Julien Benda, *Précision* (Paris: Gallimard, 1937), p. 29.

Chapter 3

1. This phrase is taken from Jean-Pierre Rioux's excellent brief discussion of the role of the intellectuals during the Algerian War, in *L'Expansion et l'impuissance*, vol. 2 of *La France de la Quatrième République* (Paris: Éditions du Seuil, 1983), p. 126. See also Michel Winock, "Les Affaires Dreyfus," *Vingtième siècle*, no. 5 (January–March 1985): 31–34.

2. Information on the second call from *NYRB*, May 12, 1966, p. 23.

3. The letter is reprinted in François Maspero, ed., *Le droit à l'insoumission: Le Dossier des 121* (Paris: Maspero, 1961), p. 227.

4. Albert Camus, "Discours à l'Académie suédoise," in *Les Prix Nobel en 1957* (Stockholm: Imprimérie Royale P. A. Norstedt & Söner, 1958), p. 48. In the deleted section Camus expatiates on the difficulty of the writer's vocation but adds, perhaps wistfully given his growing alienation from his native land, that a writer can recover a sense of a living community that will justify him. To do so he must accept the two duties that mark the nobility of his profession, "the service of truth and of liberty." Camus then reformulated this idea, but in a negative or oppositional sense that must have made it more

vivid and appealing to the sponsors of the "read-in for peace in Vietnam."

5. From an interview reprinted in Claudie Broyelle and Jacques Broyelle, *Les illusions retrouvées: Sartre a toujours raison contre Camus* (Paris: Grasset, 1982), p. 205.

6. The phrase *le colonisateur de bonne volonté* is taken from Raymond Aron's sensitive and fair-minded though ultimately critical treatment of Camus in *l'Algérie et la république* (Paris: Plon, 1958), p. 107. Aron probably borrowed the term from Albert Memmi's extremely influential *Portrait du colonisateur* (Paris: Correa, 1957). In this section, which draws a general portrait of the colonizer of good-will, Memmi does not specifically refer to Camus, but in a letter to the editors of *La Nef* he does, noting that Camus incarnates almost exactly the category he established in his book: "It is an ambiguous role, but I would insist that it is neither comical nor deserving of scorn": *La Nef* 14, no. 12 (December 1957): 95.

7. The most perceptive study in English of Camus's tormented relationship with his native land remains Conor Cruise O'Brien, *Albert Camus of Europe and Africa* (New York: Viking, 1970). Cruise O'Brien admittedly had strong ideological differences with Camus, yet he makes a convincing case that even if the fulsome praise that has been showered on Camus is merited, and his work a notable and noble expression of the "Western Moral Conscience, ... [we] should not ignore the fact that it also registers the hesitations and limitations of that conscience and that one of the great limitations lies across the cultural frontier, in the colony" (p. 28). Cruise O'Brien, himself involved in the process of African decolonization as a United Nations administrator, confronts the issue of Camus and Algeria head on, unlike Emmett Parker, who tends to be overly elegiac and discusses the Nobel Prize speech in only a paragraph. See Emmett Parker, *Albert Camus: The Artist in the Arena* (Madison: University of Wisconsin Press, 1964), pp. 150–67. Parker was either unaware of or ignored Camus's December 13, 1957, interview in Stockholm and the controversy it generated (see notes 17 and 18). For full documentation, see Herbert Lottman's massive *Albert Camus* (Garden City, N.Y.: Doubleday, 1979), esp. pp. 540–618.

8. Albert Camus, *Actuelles III, chronique algérienne 1939–1958* (Paris: Gallimard, 1958), p. 73.

9. Ibid., pp. 89–90.

10. Ibid., p. 93.

11. For a good brief discussion of this position, see John Talbott, *The War Without a Name* (New York: Knopf, 1980), pp. 19–23.

12. Camus, *Actuelles III*, p. 110.

13. Ibid., p. 122.

14. Ibid., p. 126.

15. Ibid., p. 202.

16. Ibid., p. 205.

17. See, for example, Broyelle, *Les Illusions retrouvées*, p. 195. Also, Michael Walzer, "Albert Camus's Algerian War," in his *The Company of Critics: Social Criticism and Political Commitment in the Twentieth Century* (New York: Basic Books, 1988), pp. 136–52.

18. Reported by Dominique Birmann, "Albert Camus a exposé aux étudiants suédois son attitude devant le problème algérien," *Le Monde*, December 14, 1957, p. 4. In a letter dated December 17, 1957, to the director of *Le Monde*, Camus admitted that Birmann had quoted him correctly. Extracts from Birmann's article and Camus's letter are reprinted in Albert Camus, *Essais*, ed. R. Quilliot (Paris: Gallimard, 1965), vol. 2, pp. 1881–82.

19. Birmann, in Camus, *Essais*.

20. "T. M." (editors of *Les Temps modernes*), "La Réponse d'Henri Alleg," *Les Temps modernes* 13, no. 145 (March 1958): 1530. Cf. also, Jean Conilh, "La Question et la réponse, *Esprit* 26, no. 261 (May 1958): 773–74. Conilh takes a similar position to that of *Les Temps modernes* and speaks admiringly of Sartre's preface to Alleg's book. He does offer the caution that thanks to Alleg, "the question is now inflicted upon all of us, in our souls; it will tomorrow be inflicted upon our flesh, if from this day forth we do not give an answer, if we do not speak."

21. George Braziller, Preface to Henri Alleg, *The Question*, trans. John Calder, (New York: Braziller, 1958), p. 10. Actually, similar seizures had occurred during the Vichy regime. See Anne Simonin, "Les Éditions de Minuit et les Éditions du Seuil: Deux stratégies éditoriales face à la guerre d'Algérie," in *La Guerre d'Algérie et les intellectuels français*, Institut d'histoire du temps présent, cahier no. 10 (November 1988): 151, 161–62.

22. For a detailed account in English of this formative period in Camus's life, see Patrick McCarthy, *Camus* (New York: Random House, 1982), chap. 4, "The Adventure of Alger-Républicain."

23. Quoted in Simonin, "Deux stratégies éditoriales," p. 162. See also the interview with Jérôme Lindon in Hervé Hamon and Patrick Rotman, *Les Porteurs de valises: La Résistance française à la guerre d'Algérie* (Paris: Albin Michel, 1979), p. 92.

24. For an English translation of the complete document, see

David L. Schalk, *Roger Martin du Gard: The Novelist and History* (Ithaca, N.Y.: Cornell University Press, 1967), pp. 211–12.

25. Quoted in Rioux, *L'Expansion et l'impuissance*, p. 137.

26. Camus, "Discours à l'Académie suédoise," p. 47.

27. For an excellent study in English of the Audin case, see John Talbott, "The Strange Death of Maurice Audin," *Virginia Quarterly Review* 52, no. 2 (Spring 1976): 224–42. Cf. also, Le Comité Maurice Audin, "La Mort de Maurice Audin," *Les Temps modernes* 15, no. 166 (December 1959):1118–23. This was an important communiqué, presenting the committee's findings to date. The mass-circulation press did not discuss this report in detail, although it was a very serious document, accusing officers serving in the army in Algeria of torture and murder and naming names. Curiously enough, the authorities did not publish a denial, nor did they file any sort of libel suit against the committee. By inference the latter's claims had substance, though nothing was done against the officers charged. As Maurice Péju put it in *Les Temps modernes*, the army is covering up the crime, and we have a rebellion of military power against civilian authority: "It took less than that, in the past, to unleash the Dreyfus affair": "L'Affaire Maurice Audin et la presse d'information," *Les Temps modernes* 15, no. 166 (December 1959): 1111.

28. Rioux, *L'Expansion et l'impuissance*, p. 137.

29. Ibid., p. 138.

30. "Après une soutenance symbolique à la Sorbonne, M. Maurice Audin obtint le doctorat ès sciences," *Le Monde*, December 3, 1957, p. 2; Michel Winock, *La République se meurt: Chronique 1956–1958* (Paris: Éditions du Seuil, 1978), pp. 164–66; Jean Conilh, "Thèse," *Esprit* 26, no. 257 (January 1958): 108.

Camus's ideological enemies, the French Communists, were not present either, probably because Audin, though a member of the Algerian party, did not as a European obtain authorization from the French party before joining the Algerian independence movement. Given the extreme prudence of the French Communist party during the Algerian War, it would probably have been denied had Audin asked. Just before the ceremony in Audin's honor, the perennially dissident communist intellectual, Jean Bruhat, who was in the auditorium, was enraged that none of the party leadership were in the audience. He raced to a café telephone to communicate his indignation to party headquarters, to urge them to send a delegate while there still was time. By chance he encountered the party leader Leo

Figuères who told him "not to be concerned with this affair." See Jean Bruhat, *Il n'est jamais trop tard: Souvenirs* (Paris: Albin Michel, 1983), p. 178. Also Jeannine Verdès-Leroux, "La guerre d'Algérie dans la trajectoire des intellectuels communistes," *La Guerre d'Algérie et les intellectuels français*, Institut d'histoire du temps présent, cahier no. 10 (November 1988): 215.

31. McCarthy, *Camus*, p. 284.

32. See Omar Ouzegane, as cited in Broyelle, *Les Illusions retrouvées*, p. 204. Also Frantz Fanon, "La Minorité Européenne d'Algérie en l'an V de la révolution," *Les Temps modernes* 14, nos. 156–60 (May–June 1959): 1855–63.

33. Camus, *Chronique algérienne*, p. 166.

34. See McCarthy, *Camus*, pp. 297–98, and especially Lottman, *Albert Camus*, pp. 592, 595–98, 612, 638–39. Lottman reviewed the documentation with great care, citing the lawyer Yves Déschezelles who defended Algerian prisoners and many others.

35. Birmann, in Camus, *Essais*.

36. McCarthy, *Camus*, p. 325.

37. Taken from Alice Payne Hackett, *80 Years of Best Sellers, 1895–1975* (New York: Bowker, 1977).

38. See, for example, the great admiration for Camus in Berrigan's "Dialogue Underground," *NYRB*, March 25, 1971, p. 28.

39. Francine du Plessix Gray, *Divine Disobedience: Profiles in Catholic Radicalism* (New York: Vintage Books, 1971), p. 134.

40. There was a good deal of interjournal coperation. Antiwar Catholic authors, and Protestants like Paul Ricoeur who worked closely with them, often placed articles in a variety of journals to reach the widest possible audience. *La Nef*, which published a number of moderate and thoughtful articles on Algeria by François Mitterrand, was definitely more prudent in its opposition to the war, whereas *Témoignage chrétien* was more militant. Between 1955 and 1958 *Témoignage chrétien* was seized sixty-eight times in Algeria: Renée Bédarida, "La gauche chrétienne et la guerre d'Algérie," in *La Guerre d'Algérie et les chrétiens*, Institut d'histoire du temps présent, cahier no. 9 (October 1988): 99.

41. The phrase *péché organisé par mon pays* is taken from "Fragments de vie," an antiwar poem by André Thisse, a former draftee who had just returned to mainland France after serving in Algeria. The poem was published in *Esprit* 27, no. 273 (May 1959): 753.

42. The circulation figure is from *Ulrich's International Periodicals Directory*, 23rd ed., 1984, vol. 1.

43. See Raoul Girardet, *L'idée coloniale en France de 1871 à 1962*, rev. ed. (Paris: Livre de Poche, 1979), p. 317.

44. "Une Affaire intérieure," *Esprit* 23, no. 232 (November 1955): 1646.

45. Paul Mus, *Guerre sans visage: Lettres commentées du Sous-Lieutenant Émile Mus* (Paris: Éditions du Seuil, 1961), p. 14. Selections from Paul Mus's preface appeared under the title "L'heure des vérités," in the January 1961 issue of *Esprit*, shortly before the book went to press.

46. From an interview with Jean-Marie Domenach in Rémy Reiffel, "L'Empreinte de la guerre d'Algérie sur quelques figures intellectuelles 'de gauche',," in *La Guerre d'Algérie et les intellectuels français*, Institut d'histoire du temps présent, cahier no. 10 (November 1988): 137.

47. Domenach observed that these extracurricular activities "caused us many ennuis," and it is at least possible that the government would not have harassed *Esprit* itself so severely if the editorial group had only been involved with the publication of the actual journal. See Jean-Marie Domenach, "Commentaires sur l'article de David L. Schalk," *La Revue Tocqueville* 8 (1986–87): 93–95.

48. Jean-Marie Domenach, "Génocide?" *Esprit* 29, no. 294 (April 1961): 640–42. The intellectual who made this charge was Maurice Maschino, one of the first and most articulate of the *insoumis*, the draft refusers, who went into exile in Tunisia rather than serve in Algeria.

49. See, for example, Jean-Marie Domenach, "Les Enchères de la terreur," *Esprit* 25, no. 252 (July 1957): 104–6; Jean-Marie Domenach, "Une Mauvaise Philosophie," *Esprit* 26, no. 258 (February 1958): 247–59. The April 1956 issue, which includes some strong condemnations of the war and the French policies of repression, also printed in the "Documents" section a petition signed by many Catholic intellectuals and others, protesting the existence of concentration camps in Nasser's Egypt.

50. Louis Casamayor, "Pleins pouvoirs, faux pouvoirs," *Esprit* 28, no. 282 (March 1960): 541.

51. Pierre Vidal-Naquet, "L'O.A.S. et la torture," *Esprit* 30, no. 306 (May 1962): 450–52. For other examples of the admirable fairness of Vidal-Naquet and the Comité Maurice Audin, which took on other cases besides the one that inspired its founding, see the committee's report on the evidence that had come to its attention concerning the torture of suspected members of the OAS, in *Le Monde*, October 18, 1961, p. 16.

52. See especially Paul Ricoeur, "Le 'cas' Etienne Mathiot," *Esprit* 26, no. 259 (March 1958): 450–52.

53. See, for example, Paul Ricoeur, "L'Insoumission," *Esprit* 28, no. 288 (October 1960): 1600–4; Jean Conilh, "La voix de la France," *Esprit* 28, no. 289 (November 1960): 1965–67.

54. Among the many articles, see especially Paul Thibaud, "Question de conscience," *Esprit* 27, no. 271 (March 1959): 476–77; and the unsigned article, the longest of any printed during the Algerian War, "Histoire d'un acte responsable, le cas Jean Le Meur," *Esprit* 27, no. 279 (December 1959): 675–707. Le Meur was a young second lieutenant who had been in combat and refused further service after he heard the commanding officer repeat, "I do not want any prisoners." On alternative service, see Jacques Tinel, "Construire des choses solides et vrais, *Esprit* 28, no. 260 (December 1960): 2061–64; Henri Bartoli, Alain Rouzet, Emile Galey, Alain Zarudiansky, "Leur acte nous engage," *Esprit* 28, no. 290 (December 1960): 2064–66.

55. Jacques Tinel, "X2: Se disant Jack Muir," *Esprit* 29, no. 293 (March 1961): 464–67.

56. See, for example, Max Milner, "Une Conscience en correctionnnelle," *Esprit* 29, no. 298 (September 1961): 266–70. Milner reported on the trial in Dijon of a draft resister named Michel Halliez. The defense lawyer was Robert Badinter, who was minister of justice in the Mitterrand government between 1981 and 1986.

57. For a discussion of Mounier's role in the formulation and diffusion of the concept of engagement, see David L. Schalk, *The Spectrum of Political Engagement: Mournier, Benda, Nizan, Brasillach, Sartre* (Princeton, N. J.: Princeton University Press, 1979), pp. 17–25.

58. Jean-Marie Domenach, "Est-ce que la guerre d'Afrique du Nord?" *Esprit* 22, no. 221 (December 1954): pp. 768–70. Domenach was working from a tradition here, in that *Esprit* under under its founder Emmanuel Mounier had been deeply concerned with colonialism. Mounier ran a series of articles in the July 1947 issue, entitled "To Prevent War in North Africa."

59. Ibid., p. 770. One cannot help but be struck by how quickly Domenach denounced torture in Algeria.

60. Ibid.

61. Colette Jeanson, "L'Algérie à la 'une'," *Esprit* 23, no. 228 (July 1955): 1248.

62. François Sarrazin, "l'Algérie, pays sans loi...," *Esprit* 23, nos. 230–31 (September–October 1955): 1624, 1629–30, (Italics his)

63. François Sarrazin, "L'Afrique du Nord et notre destin," *Esprit* 23, no. 232 (November 1955): 1643, 1665.

64. "Esprit" (the editors jointly), "Une affaire intérieure," *Esprit* 23, no. 232 (November 1955): 1642, 1645–46.

65. Evaluating the legitimacy of that claim is not central to our concerns here. Only in the 1980s was it seriously questioned. The case of the founder of *Esprit*, Emmanuel Mounier, who wavered briefly during the Occupation but was soon firmly and courageously in the Resistance camp, has been thoughtfully evaluated by John W. Hellman in "Emmanuel Mounier: A Catholic Revolutionary at Vichy," *Journal of Contemporary History* 8, no. 4 (October 1973): 3–23. Hellman has also examined the scandal that erupted in 1981, when Bernard-Henri Lévy published *L'Idéologie française*. Lévy charged that several prominent Catholic intellectuals, normally placed at the left of the political spectrum and associated with the Resistance, were in truth fascists of a certain variety, implicated in a "fascism with French colors." Cf. Hellman's "Personalisme et fascisme," in Association des Amis d'Emmanuel Mounier, ed., *Le Personnalisme d'Emmanuel Mounier*, pp. 116–42 (Paris: Éditions du Seuil, 1985).

66. See Georges-E. Lavau, "Au-delà de la violence et de la honte," *Esprit* 23, no. 232 (November 1955): 1697. If I were an Algerian, Lavau writes, I would find Soustelle's plan to "reform" Algeria by creating two more departments for a total of five, as the "last slap in the face that would leave me no other recourse than revolt. As a Frenchman, I am overwhelmed with shame." Some of the most moving passages of Simone de Beauvoir's memoirs for these years recount her sense of shame at her ineradicable Frenchness, as she observes the racism of her compatriots and their near-unanimous support of the war in its early stages. See *La Force des choses* (Paris: Gallimard, 1963), vol. 2, pp. 145–48, 239, 396, 453. See Also Henri Marrou's powerful editorial, which had a major impact leading to counter editorials and to a police search of his apartment: "France, ma Patrie," *Le Monde*, April 5, 1956, p. 2. In responding to his critics on the right, François Mauriac put the issue of patriotism as it affected moderate antiwar intellectuals as powerfully as any writer: " 'You never denounce the crimes of the others.' I have said and I repeat it. Do you not think that I am horrified by them? But we are not accountable for them. It does not depend on us French that they be perpetrated or not, even if they are the fruit of our policies. It is not in our name that they are carried out": François Mauriac, *Le Nouveau Bloc-Notes, 1958–1960* (Paris: Flammarion, 1961), p. 29.

67. Jean-Marie Domenach, "Une nouvelle opinion catholique,

Esprit 23, no. 232 (November 1955): 1768. On Duval, see Georges Houdin, "Ce que fut l'attitude des chrétiens français," *La Nef*, cahier nos. 12–13 (October 1962–January 1963): 73.

68. Girardet, *l'Idée coloniale en France*, p. 383. Information on Monseigneur Rodhain from Jean-Marie Domenach, "Le Prélat, l'objecteur...," *Esprit* 28, no. 281 (February 1960): 286.

69. Domenach, "Une nouvelle opinion catholique," p. 1770.

70. André Nozier, "La communauté catholique d'Algérie et la guerre," in *La Guerre d'Algérie et les chrétiens*, Institut d'histoire du temps présent, cahier no. 9 (October 1958): 11–12. Nozier had access to unpublished materials pertaining to the church's role during the Algerian War.

71. See, for example, Jean-Marie Domenach, "Négocier en Algérie," *Esprit* 24, no. 236 (March 1956): 322. "This war cannot be won. They can only prevent its being lost."

72. On Nizan, who died in 1940 but who became a kind of guiding spirit for the French resistance to the Algerian War, see my *Spectrum of Political Engagement*, esp. pp. 52–53, 136. One of the clandestine student groups supporting Algerian independence called itself the "Paul Nizan cell."

73. Jean-Marie Domenach, "Lettres d'Algérie," *Esprit* 24, no. 235 (February 1956): 250.

74. Jean Sénac, "Lettre à un jeune français d'Algérie," *Esprit* 24, no. 236 (March 1956): 336.

75. Alain Berger, "La République en danger," *Esprit* 24, no. 237 (April 1956): 577.

76. See Michel-Antoine Burnier, *Les Existentialistes et la politique* (Paris: Gallimard, 1966), pp. 114–48; also David Caute, *Communism and the French Intellectuals, 1914–1960* (New York: Macmillan, 1964), pp. 255–57.

77. Jean-Marie Domenach, "Notre faute," *Esprit* 24, no. 245 (December 1956): 895.

78. Jean-Marie Domenach and Jacques Julliard, "Réveiller la France," *Esprit* 25, no. 246 (January 1957): 78.

79. Ibid., p. 79.

80. Jean-Marie Domenach, "Démoralisation de la nation," *Esprit* 25, no. 249 (April 1957): 577–79. Domenach recalled that he went to see an officer in the War Ministry to show him the documents and tell him that he would publish them if the atrocities did not cease. "He treated me like a 'boy scout,' and I published": "Commentaires sur l'article de David L. Schalk," p. 93.

81. Jean-Marie Domenach, "Algérie, propositions raisonnables," *Esprit* 25, no. 250 (May 1957): 778–79.

82. As discussed in Chapter 2.

83. Years later Domenach was more critical of Jeanson than he had been in his articles in *Esprit* and public statements during the war, when he testified on behalf of the arrested members of the "Jeanson ring" during their famous 1960 trial. Jeanson's group followed an "aberrant and dangerous political line," and their political analysis was "puerile," Domenach wrote in 1987. Domenach recalled that they had a conversation while Jeanson was in hiding, and Jeanson announced to him that "insurrection" was about to break out in France: "Commentaires sur l'article de David L. Schalk," p. 94.

84. Yves Goussault, "Manifestations silencieuses," *Esprit* 25, no. 253 (September 1957): 250. Information on the details of the demonstration and the number arrested is from *Le Monde*, June 25, 1957, p. 2.

85. Ibid., pp. 250–51.

86. Paul Ricoeur, "Le 'cas' Etienne Mathiot," *Esprit* 26, no. 259 (March 1958): 451.

87. Robert O. Paxton, *Vichy France: Old Guard and New Order, 1940–1944* (New York: Knopf, 1972), p. 383.

88. Ricoeur, "Le 'cas' Etienne Mathiot," p. 452.

89. Daniel Berrigan and Robert Coles, "A Dialogue Underground," *NYRB*, March 11, 1971, p. 19; Daniel Berrigan, *Absurd Convictions, Modest Hopes* (New York: Vintage Books, 1973), pp. 91, 203–5, 222–27.

90. In addition to the articles cited in notes 53 through 56, see Jean-Marie Domenach, "Le Procès des non-violents," *Esprit* 30, no. 302 (January 1962): 100–3.

91. Jean-Marie Domenach, "Introduction to 'Histoire d'un acte responsable, le cas Jean Le Meur,'" *Esprit* 27, no. 279 (December 1959): 676.

92. My private files (I was head of the Dutchess Country Draft Counseling and Information Service from 1968 to 1970). In an important article entitled "Insoumission" (Insubordination), Paul Ricoeur clearly draws the distinction between informational and political draft counseling as it applied to the Algerian War. He explains that he will not counsel insubordination but refuses to condemn it, and if ordered he will state his reasons before a military tribunal. This article was widely circulated, first published in *Cité*

nouvelle and reprinted in *Esprit* 28, no. 288 (October 1960): 1600–4. It also was published in Maspero, ed., *Le droit à l'insoumission*, pp. 143–47.

93. Jean-Marie Domenach, "Sauve-qui-peut?" *Esprit* 28, no. 283 (April 1960): 707. For an account in English of the activities of the Jeanson network, see Paul C. Sorum, "The New Resistance" (chap. 6), in his *Intellectuals and Decolonization in France* (Chapel Hill: University of North Carolina Press, 1977).

94. Domenach, "Sauve-qui-peut?" p. 708.

95. Jean-Marie Domenach, "Résistances," *Esprit* 28, no. 284 (May 1960): 804.

96. Ibid., p. 805. Among the many studies of the fall of the Fourth Republic, see especially Charles D. Maier and Dan S. White, *The Thirteenth of May: The Advent of de Gaulle's Republic* (New York: Oxford University Press, 1968).

97. Domenach, "Résistances," p. 807. Here and elsewhere in analyses of and appeals for engagement written during the Algerian War, one finds the metaphor of the "obstructed path," used so tellingly by H. Stuart Hughes in *The Obstructed Path: French Social Thought in the Years of Desperation, 1930–1960* (New York: Harper & Row, 1968).

98. Domenach, "Résistances," p. 808.

99. See Domenach's bitter article, "La Nature des choses," *Esprit* 28, no. 288 (October 1960): 1599. Here Domenach admits that the recent trials of the members of the Jeanson network whom the police had managed to catch, and desertion and insubordination in the army, were representative of only a small minority. Yet these kinds of actions were symptomatic of a "moral fissure, the beginning of a secession that will increase with the continued pursuit of the war. In the months ahead, those who consider themselves still to be citizens are going to be obliged to assume their responsibilities, and the government will not for long escape its own."

100. Conveniently reprinted along with a list of the signatories in Maspero, ed., *Le Droit à l'insoumission*, pp. 15–18. The text is also available in Hamon and Rotman, *Les Porteurs de valises*, pp. 391–94.

101. Reprinted in Maspero, ed., *Le Droit à l'insoumission*, pp. 155–57.

102. Jean Conilh, "La Voix de la France," *Esprit* 28, no. 289 (November 1960): 1965–66.

103. Jean-Marie Domenach, "La Manif.," *Esprit* 30, no. 303 (February 1962): 248–49.

104. When the demonstration finally took place, the reports in *Esprit*, which are powerfully and evocatively written, express a tone of surprise, almost shock. "We have never seen anything like it": Louis Casamayor, "Pour la liberté," *Esprit* 30, no. 304 (March 1962): 427. It represented the "authentic republican tradition, which escapes the logic of political reasoning, one of those privileged moments when the energy and spiritual vitality of a people who had been written off as dormant erupted into history": Jean Conilh, "Retour aux sources," *Esprit* 30, no. 304 (March 1962): 431–32.

105. Le Comité directeur d'*Esprit*, "Contre la barbarie," *Esprit* 29, no. 300 (November 1961): 669–70. See also Philippe Ivernel's powerful appeal, "Vaincre la ségrégation," *Esprit* 29, no. 301 (December 1961): 907. Ivernel reminds his readers that the whereabouts of the thousands of Algerians arrested for peacefully demonstrating on the streets of Paris—away from their "ghetto"—was widely known. They had been locked up in two sports arenas where they were brutally treated. "And where was the Communist party, one for all, all for one? Where was [Pierre Mendès-France's] Unified Socialist party, the party of peace?...Where were the few of us from *Esprit*?"

106. See, for example, "Réussir la paix," *Esprit* 30, no. 305 (April 1962): 700–4. (A position paper representing the views of the entire editorial committee)

107. The most careful and convincing examination of these estimates was made by Guy Pervillé. See his "Bilan de la guerre d'Algérie," in *Études sur la France de 1939 à nos jours* (special issue of *L'Histoire*, Paris: Éditions du Seuil, 1985), pp. 297–301. Sartre makes his claim in his preface to Frantz Fanon, *Les Damnés de la terre* (Paris: Maspero, 1961), p. 23.

108. Philippe Ivernel, "Paris-Match à l'heure du cessez-le feu," *Esprit* 30, no. 307 (June 1962): 980.

109. Ibid., p. 981.

110. Talbott, *The War Without a Name*, p. 249.

111. Quoted in Reiffel, "L'Empreinte de la guerre d'Algérie sur quelques figures intellectuelles," p. 144.

112. Editorial note, *Esprit* 26, no. 265 (September 1958): 193.

113. Jean-Marie Domenach, "L'Après-guerre," *Esprit* 30, no. 310 (October 1962): 353. Already in November 1962, Pierre Vidal-Naquet felt obliged to remind the French people that "although we tend today to forget it, the Algerian War did take place." The citation is from a moving and courageous article in *Le*

Monde denoncing the new Algerian regime for its murderous policy toward the *harkis*, native Algerians who had served as irregulars in the French army, only a small number of whom, for economic and other reasons, were allowed to escape to France in the summer of 1962. The *harkis*, of whom more than 100,000 were recruited, sometimes by force, should not have to pay for the faults of France, and a major effort should be made to save them and their families: "La Guerre révolutionnaire et la tragédie des harkis," *Le Monde*, November 11–12, 1962, p. 11.

114. I. F. Stone, "The Hidden Traps in Nixon's Peace Plan," *NYRB*, March 9, 1972, p. 16.

115. Because historians are also citizens, they may believe that in some cases, when the social fabric would be damaged by remembering, certain past events are best left forgotten. Where the universal agreement among historians comes into play is that we all concur that whatever subjects we have chosen to study deserve to be remembered.

116. In 1987 Domenach wrote that the various kinds of governmental harassment that the *Esprit* group endured during the Algerian War were really not serious for people like himself who had known the armed Resistance. Nonetheless the latter period was "much more painful for me. We were not combating an enemy, but our government. During a year in the *maquis*, I had dreamed of wearing this uniform that now I saw dirtied by torture": "Commentaires sur l'article de David L. Schalk," p. 94.

117. Jean-Marie Domenach, *Ce que je crois* (Paris: Grasset, 1978), p. 186.

118. Pierre Vidal-Naquet, "Une Fidélité têtue: La Résistance française à la guerre d'Algérie," *Vingtième siècle*, no. 10 (April–June 1986): 11.

119. Ibid.

120. Robert Bonnaud, "Le Vrai crime," *Esprit* 25, no. 251 (June 1957): 1007.

121. Ali Haroun, *La 7e Wilaya. La Guerre du FLN en France, 1954–1962* (Paris: Éditions du Seuil, 1986), p. 109.

122. Ibid., pp. 307, 313.

123. Vidal-Naquet, "Une Fidélité têtue," p. 12.

124. Annie Cohen-Solal, *Sartre* (Paris: Gallimard, 1985), pp. 555–56.

125. Jean-Paul Sartre, "Le Colonialisme est un système," *Les Temps modernes* 11, no. 123 (March–April 1956): 1371–86. Later reprinted in *Situations V* (Paris: Gallimard, 1964), pp. 25–48. Sartre

argued that there is no such thing as a good colonialist and that reforms will not work. The reformation of Algeria will be the responsibility of the Algerian people themselves, "when they have conquered their liberty." In fact he believes that Algerian liberation and "that of France can only emerge upon the destruction of colonization" (p. 1372).

126. The brief text reads as follows:

"Two months after the cease-fire, several dozen Frenchmen remain in prison on charges of aiding Algerian militants or for refusing to fight; others have been forced into exile.

"Their motives were diverse, and we have divergent appreciations of their actions. But we all believe that at a moment when cooperation between France and the new Algeria is recognized as a national task and an imperative for the future, the place of these men is not in prison or abroad. Draft resisters or members of 'Networks of Aid for the FLN,' they are awaiting the decision that will permit them to devote themselves, now that we are at peace, to the ideal they wished to serve during wartime.

"We demand thus that a rapid and complete amnesty permit these draft resisters and these prisoners to return to their normal lives as citizens": "Pour l'Amnistie," *Esprit* 30, no. 308 (July–August 1962): 191–92.

127. Jean-Paul Sartre, "Une Victoire," originally published in *L'Express*, March 6, 1958, reprinted in *Situations V: Colonialisme et néocolonialisme* (Paris: Gallimard, 1964), pp. 72–88. The citation is from p. 75. In this piece Sartre refers to 1943 and the Nazi tortures in Paris; at that point it seemed impossible "that we could one day make men cry out in pain in our name." However, *"impossible n'est pas français"* (p. 72).

128. Jean Rous, "Nationalisme et révolution d'outremer," *Les Temps modernes* 10, nos. 112–13 (May 1956): 1954–55. This was a special double issue dedicated to the Left. Rous's article attempted to cover in twelve pages what was becoming known as the Third World. The brief section on Algeria shows much less awareness than did Domenach six months earlier, hardly mentioning that a war was going on. Rous argues against separatism (i.e., independence) but, rather, favors the application of genuine democracy. Because the laws guaranteeing free elections in Algeria had been violated by the colonial administration and basic liberties were denied, the situation had become desperate, and the "recent explosion in the Aurès [mountains] has occurred."

129. The question is not addressed in Michel-Antoine Burnier's

otherwise useful *Les Existentialistes et la politique* (Paris: Gallimard, 1966), translated as *Choice of Action* (New York: Random House, 1969). The Broyelles might have used it for ammunition in their attack on Sartre in *Les Illusions retrouvées,* but they do not.

130. *LTM* 22, no. 118 (October 1955): 385–88. This editorial is quite a dramatic jump into the fray of discussion, especially because it is the first detailed commentary on the war. The editors fear that the government is deliberately moving toward catastrophe and dismiss as an idiotic fiction the claim that Algeria—an exploited colony in their view—is composed of three French departments. They denounce torture, repression, arbitrary arrest, and electoral fraud. The authorities cannot dissimulate the truth that the "rebels" or "outlaws" are in reality "soldiers in a Resistance movement that an entire people aids and protects." Although they are not yet explicit, they already state indirectly that they would support draftees who refuse to serve in Algeria.

131. Which the editors of *LTM* had denounced in an editorial, "Pouvoirs spéciaux," *LTM* 11, no. 1234 (March–April 1956): 1345–53.

132. Annie Cohen-Solal, *Sartre* (Paris: Gallimard), 1985, p. 563 (citing a 1984 interview with the lawyer Roland Dumas, who had been the chief counsel for the Jeanson network). Cf. Jean-François Sirinelli, "Guerre d'Algérie, guerre des pétitions?" *La Guerre d'Algérie et les intellectuels français,* Institut d'histoire du temps présent, cahier no. 10 (November 1988): 181, 200.

133. "Le Colonialisme est un système" and "Une Victoire" were cited in notes 125 and 127. "Vous êtes formidables," which appeared in the May 1957 issue of *LTM* and was reprinted in *Situations V,* pp. 57–67, was chosen by Cohen-Solal as the title for one of her chapters. It is first a powerful moral denunciation of French methods of pacification, inspired by the publication of a collection of depositions and documents on "that gangrene," the use of torture and murder by the army in Algeria. Then there is a sardonic look at a popular radio personality who occupied himself with stroking the French national ego, incessantly telling the French that they were "formidables." We French are neither formidable nor candid, Sartre argues, because we hide from ourselves the truth of what is occurring in Algeria.

134. "Nous sommes tous les assassins" was first published in *LTM* in March 1958 and reprinted in *Situations V,* pp. 68–71.

135. Maurice Maschino, "Le Refus," *LTM* 14, no. 152 (October 1958): 701–16. Maschino published numerous articles and interviews

in *LTM*, including "Le Dossier des réfractaires," in vol. 15, no. 169–70 (April–May 1960): 1550–62. This document dealt with the "escape network" that helped young draft evaders get out of the country. Many of Maschino's writings on the war and draft resistance were collected into a volume and published by the *tiersmondiste* editor François Maspero under the tile of *L'Engagement* (Paris: Cahiers libres, no. 19, 1961).

136. Francis Jeanson, "Lettre à Jean-Paul Sartre," *Les Temps modernes* 15, no. 169–70 (April–May 1959): 1536. Jeanson cites three principal reasons for his engagement: first, to keep alive the chances of Franco-Algerian friendship; second, to revive the French Left; and third and interestingly, "to save the honor of France and her most valid traditions," at a time when France was heavily criticized by world opinion for her repression in Algeria. This is yet another indication that even the most extreme forms of engagement during the Algerian War were at least in part motivated by a sense of wounded patriotism.

137. Ibid., pp. 1538, 1543.

138. "Le Premier Congrès de 'Jeune Résistance,' " *LTM* 16, nos. 173–74 (August–September 1960): 312–28.

139. Jacques Vergès, La Leçon d'un procés," *LTM* 16, nos. 175–76 (October–November 1960): 517. Cf. note 97.

140. Cohen-Solal, *Sartre*, p. 503.

141. Ibid., p. 504.

142. For details, see Claude Liauzu, "Intellectuels du Tiers Monde et intellectuels français: Les années algériennes," in *La Guerre d'Algérie et les intellectuels français*, Institut d'histoire du temps présent, cahier no. 10 (November 1988): 107. After the Algerian War ended, the complete manifesto finally circulated legally. See note 100.

143. Jacques Fauvet, "Les Sanctions contre les artistes et fonctionnaires signataires du manifeste sur l'insoumission," *Le Monde*, September 30, 1960, p. 1. The list of signers is given on page 6 of the same issue. The original drafters of the manifesto were not seeking a great number of signatures but were concerned with a shock effect, and the names of Clara and Florence Malraux were certainly embarrassing to the government. The initial solicitation was halted randomly at 121, because the number appealed to the organizers.

144. Given in *Le Monde*, October 7, 1960, p. 1. Also reprinted in Maspero, ed., *Le droit à l'insoumission*, pp. 175–77. This document, officially known as the "Manifesto of the French Intellectuals," is usually identified with Marshal Alphonse Juin, its best-known sig-

natory. Juin was France's most respected soldier after General de Gaulle and was a *pied-noir* and a member of the French Academy.

145. The letter, which was read by Maître Roland Dumas during the Jeanson trial on September 20, 1960, was not composed by Sartre, as he was in Brazil at the time, and his signature was forged. Sartre later fully endorsed it, and it is reprinted in Maspero, ed., *Le Droit à l'insoumission*, pp. 85–88.

146. On Sartre's militancy during this period, see Cohen-Solal, *Sartre*, p. 547; and Marie-Christine Granjon, "Raymond Aron, Jean-Paul Sartre et le conflit algérien," in *La Guerre d'Algérie de les intellectuels français*, Institut d'histoire du temps présent, cahier no. 10 (November 1988): 79–94. Also Liauzu, "Intellectuels du Tiers Monde et intellectuels français," pp. 105–18.

Among the many treatments in English, see the largely critical discussion in Hughes, *The Obstructed Path*. Hughes, however, does argue that at the time of the Manifesto of the 121, "Sartre came closest to political greatness as he voiced the shame and anger of professors and writers, of pastors and priests, revolted by the tortures and barbarities that France's war of repression had entailed" (p. 240). Although I have used the French version here, Cohen-Solal's biography is available in English (New York: Pantheon, 1985). Sartre figures extensively in Sorum's *Intellectuals and Decolonization in France*.

147. See Cohen-Solal, *Sartre*, pp. 547–48, 562–63. Among those joining the call was Jean-Marie Le Pen, at the time a recently discharged paratroop lieutenant and member of the Chamber of Deputies, who later became the head of France's extreme-right Front National.

148. Jean-Paul Sartre, "Les Somnambules," *LTM* 17, no. 191 (April 1962). 1397–1401. Also published in *Situations V*, pp. 160–166 (incorrectly dated February 19, 1962, in both sources).

149. Jean-Pierre Azema, Jean-Pierre Rioux, and Henry Rousso, "Les Guerres franco-françaises," *Vingtième siècle*, no. 5 (January–March 1985). 4. Indeed, there is some uncertainty as to whether the Algerian War ever acquired the status of a true "Franco-French war," as the Vichy episode definitively did.

150. Talbott, *The War Without a Name*, p. 249.

151. Michel Crouzet, "La Bataille des intellectuels français," *La Nef*, cahier nos. 12–13 (October 1962–January 1963): 47, 50–52.

152. Ibid., p. 47.

153. Bernard Droz, "Le Cas très singulier de la guerre d'Algérie," *Vingtième siècle*, no. 5 (January–March 1985): 89. Exactly the

same language could be applied to the American intellectual class and Vietnam.

154. Jean-Paul Sartre, "Sartre on the Nobel Prize, *NYRB*, December 17, 1964, pp. 5–6.

Chapter 4

1. Quoted by Susan Sontag, in Cecil Woolf and John Bagguley, eds., *Authors Take Sides on Vietnam* (New York: Simon & Schuster, 1967), p. 70.

2. See David J. Armor, Joseph P. Giacquinta, R. Gordon McIntosh, Diana E. H. Russell, "Professors' Attitudes Toward the Vietnam War," *Public Opinion Quarterly* 31, no. 2 (Summer 1967): 159–75. Their sample consisted of 152 professors, and their interviews were carried out in April and May 1966, early in the war. Also see Howard Schuman and Edward O. Laumann, "Do Most Professors *Support* the War?" *Trans-action* 5, no. 1 (November 1967): 32–35. The interpretation of the data they drew from a survey taken at the University of Michigan is, in my view, suspect. In any case they relied on 242 questionnaires, whereas 608 faculty members, one fifth of the total, had signed a public letter to President Johnson calling for an unconditional halt to the bombing raids on North Vietnam.

3. E. M. Schreiber, "Opposition to the Vietnam War Among American University Students and Faculty," *British Journal of Sociology* 24, no. 3 (September 1973): 293.

4. See Melvin Small, *Johnson, Nixon and the Doves* (New Brunswick, N.J.: Rutgers University Press, 1988), pp. 183–84.

5. For a discussion of this incident see ibid., p. 119. The aide was Robert Ginsburgh.

6. Woolf and Bagguley, *Authors Take Sides*, p. 15. In yet another reference to the terrible memory of the Spanish civil war, which influenced antiwar intellectuals in both France and the United States, Woolf and Bagguley dedicated their work to Nancy Cunard, who had conceived and compiled *Authors Take Sides on the Spanish Civil War* thirty years earlier.

7. For Vidal, "should the war in Vietnam continue after the 1968 election, a change in nationality will be the only moral response" (ibid., p. 73). Kay Boyle, who had visited Cambodia in 1966, felt that our involvement in Vietnam was "the most shameful page and the most concerted record of immorality in American history" (p. 24).

8. Ibid., p. 22.

9. Ibid., p. 23. As Frederic Raphael put it, "In the end the only reasonable comparison is with Algeria, from which even the King of the Ostriches was obliged to withdraw" (p. 63).

10. Charles Kadushin, *The American Intellectual Elite* (Boston: Little, Brown, 1974), p. 124. This was an increase from Kadushin's estimate that at the end of 1965, 75 percent of America's elite intellectuals were opposed to the war (p. 133).

11. W. H. Auden, "Saying No," *NYRB*, July 1, 1971, p. 41.

12. Pascal Ory and Jean-François Sirinelli, *Les Intellectuels en France, de l'affaire Dreyfus à nos jours* (Paris: Armand Colin, 1986), p. 20.

13. The phrase is taken from Jean-François Sirinelli, "Guerre d'Algérie, guerre des pétitions? Quelques jalons," in *La Guerre d'Algérie et les intellectuels français,* Institut d'histoire du temps présent, cahier no. 10 (November 1988): 181–210. This openly admitted borrowing from the Manifesto of the 121 by the Americans is discussed in Chapter 2.

14. Everett Carll Ladd, "American University Teachers and Opposition to the Vietnam War," *Minerva* 8 (1970): 543. Ladd goes on to note that in the Sunday *New York Times* alone between October 1964 and June 1968, more than twenty thousand individuals affiliated with universities signed one or more of the antiwar petitions published in its pages.

15. Elizabeth Fox Genovese and Eugene D. Genovese, "The National Petition Campaign," *NYRB*, June 4, 1970, pp. 59–60.

16. Noam Chomsky, *American Power and the New Mandarins* (New York: Vintage Books, 1969), p. 8.

17. "Scientists and engineers" were loosely defined. Humanists could join if they were on the MIT faculty, as I was at the time.

18. "International Conference on Alternative Perspectives on Vietnam," *NYRB*, September 16, 1965, p. 11; *New Republic*, September 11, 1965, pp. 22–23. This petition included only a partial list of 224 sponsors, among whom were the Nobel Prize–winning scientists Hans Bethe and Linus Pauling, the cartoonist Jules Feiffer, the theologian Harvey Cox, the poets Stanley Kunitz, Denise Levertov, and Robert Lowell, the sociologists David Riesman and Talcott Parsons, the political scientist Hans Morgenthau, the botanist and presidential candidate in 1980, Barry Commoner, and Noam Chomsky.

19. One further illustration of the rapid escalation of opposition to the war is that by June 5, 1966, more than 6,400 academics and other professionals could sign an antiwar text published in the

New York Sunday Times, their names filling three pages of the "News of the Week in Review."

20. Published in *Ramparts* 5, no. 12 (June 1967) and reprinted six times in that magazine through July 1970. Also in the *New Republic* November 18, 1967; *The Nation*, November 6, 1967; and in the *NYRB*, December 7, 1967, p. 7. An exact facsimile was reprinted in the *NYRB*, June 5, 1969, p. 15. A prefatory note was added, stating that 100,000 people had signed this declaration during the Johnson administration. "Nixon is allowing death and destruction to continue and to increase. Pressure and action against the war are needed."

21. The "Call" differed from its French forerunner in several ways. First, there was no governmental censorship to prevent its publication, and it continued to circulate and gather signatures until the summer of 1968, when the total was over four thousand. There were also certain differences in the content of the texts. The signers of the American call believed that their statement was covered by the free speech provisions of the First Amendment and that the antiwar actions they intended to undertake were "as legal as is the war resistance of the young men themselves." It is clearly implied, though not admitted, that these projected actions would be technically illegal according to legislation in force within the boundaries of the United States, that is, counterlegal in the sense in which we have used that phrase.

22. Although obviously only men were subject to the draft and assigned combat roles in Vietnam, from the perspective of 1990 the completely male focus of the language of all the important petitions is striking. It did not seem to occur to the creators of these documents that many "women of goodwill" were as opposed to the Vietnam War as they were. Observers and analysts of the revival of American feminism, which began in the late 1960s, have often commented on the disillusion of female antiwar activists as playing a major role in launching the women's movement. Women in the peace movement were almost universally assigned only the most menial tasks by the male leadership.

23. As Melvin Small points out, petitions eventually lost their newsworthiness and their ability to shock. By 1969, an antiwar petition circulated by Dwight Macdonald, which gathered nine thousand signatures, was reported only on page 5 of the April 13 issue of the *New York Times*: *Johnson, Nixon, and the Doves*, p. 175.

24. "If a thousand men were not to pay their tax bills this year" *NYRB*, February 15, 1968, p. 9. Also published in *Ramparts* 6, no. 7 (February 1968): 60–61.

25. "Petition for Redress of Grievances," *NYRB*, June 15, 1972, p. 37.

26. From a speech published in *Le Monde*, July 9, 1957, p. 4.

27. John Talbott, *The War Without a Name: France in Algeria, 1954–1962* (New York: Knopf, 1980), p. 112.

28. Small, *Johnson, Nixon, and the Doves*, pp. 18–19.

29. Ibid., p. 105.

30. Kadushin, *The American Intellectual Elite*, p. 154.

31. *New York Times*, April 21 and 25, 1967, front page articles.

32. Talbott, *The War Without a Name*, p. 113.

33. The short and passionate life of *Ramparts* is perhaps unique in the history of American magazine journalism. Based in San Francisco, it began in May 1962 under the editorship of Edward M. Keating as the *National Catholic Journal*, with a press run of a few hundred copies. It was an elegant and rather prudish quarterly, highly intellectual, publishing essays by authors like Gabriel Marcel, poetry, art, and fiction, printed on beautiful stock paper. Then, under the impact of Vietnam it changed its format and tone, became a monthly and briefly a bimonthly, accepted advertisements, and began to publish remarkable exposés dealing with the war. As it became a kind of radical news magazine, its intellectual focus was abandoned, and its Catholic orientation almost disappeared. Politically it shifted dramatically, moving very far leftward. It began to attack Cardinal Francis Spellman and then President Johnson for their roles in Vietnam, publishing large excerpts from Barbara Garson's *MacBird!* in December 1966. It also introduced such authors as Eldridge Cleaver. At its peak in 1967, its press run was 300,000. Probably its most famous issue appeared in January 1967 with a crucified soldier featured on the cover. This issue, which was used effectively in antiwar rallies included perhaps the most painful and morally distressing article ever published on the war, "The Children of Vietnam." The preface was by Dr. Benjamin Spock, and the graphic color photographs of the horribly wounded and mutilated children were unforgettable. They must have pushed many undecided readers into the antiwar camp.

By 1968 *Ramparts* had traveled to the left edge of American politics, cooperating with the Black Panther movement. After 1968 its readership began to drop, and it flirted on several occasions with bankruptcy. Toward the end of its life its editors changed frequently, and it survived the Vietnam War by only a few months, with the last issue appearing in August 1975.

34. Sandy Vogelgesang based her pioneering study of changing

intellectual perceptions of the conflict in Indochina primarily on
material from the *NYRB*, the *Partisan Review*, the *New Republic*, and
Studies on the Left. She found the *New Republic* somewhat more polite
and moderate than the other three in its expression of antiwar dissent. See *The Long Dark Night of the Soul* (New York: Harper & Row,
1974), pp. 4, 93–94. Vogelgesang makes a good case for this assess-
ment, though she neglects to point out that the *New Republic* was
willing to take the risk of publishing the "Call to Resist Illegitimate
Authority" in 1967.

35. Kadushin, *The American Intellectual Elite*, p. 53. See also
Philip Nobile, *Intellectual Skywriting: Literary Politics and the New York
Review of Books* (New York: Charterhouse, 1974), chap. 4, "The In-
telligentsia at War."

36. Tom Wicker, "The Malaise Beyond Dissent," *New York
Times*, March 12, 1967, p. E-13.

37. Thich Nhat Hanh, "A Buddhist Poet in Vietnam," *NYRB*,
June 9, 1966, pp. 4–5.

38. Kadushin, *The American Intellectual Elite*, p. 47. The influential intellectuals that Kadushin interviewed thought that there was
only "one organ for intellectuals—the *NYRB*" (p. 44). See also p. 50
on its high ratings and p. 52 on the journal's power to make and
break reputations.

39. Nobile, *Intellectual Skywriting*, p. 5. James M. Naughton,
"Agnew, in Delaware, Criticizes 'Elitism,'" *New York Times*, October
15, 1970, p. 52.

40. Again, to keep the record straight it is important to note
that the major article in that issue (August 24, 1967) was Tom Hay-
den's reportage on the Newark riots; that is, the reference intended
by the cover was to urban violence in the ghetto, not specifically to
Vietnam protest.

41. "Protest," *NYRB*, March 14, 1968, pp. 36–37. The four
Spock trial defendants were William Sloan Coffin, Michael Ferber,
Mitchell Goodman, and Marcus Raskin, who it will be recalled, was
one of the drafters of the "Call to Resist Illegitimate Authority." His
codrafter, Arthur I. Waskow, also signed it. The other two petitions
are "An Appeal to the U.N. Committee on Human Rights," *NYRB*,
August 21, 1969, p. 37; and Joan Baez et al., "An Open Letter to
the Members of the Communist Party of the Soviet Union," *NYRB*,
June 28, 1973, p. 7. Among the signers of this statement, protesting
the persecution of men and women for their political views, were
Philip Berrigan, Ramsay Clark, Erich Fromm, Norman Mailer, Gun-
nar Myrdal, Noam Chomsky, and, interestingly enough, Arthur

Schlesinger, Jr. Chomsky had savaged Schlesinger as a confessed liar and an amoral sycophant in one of the most dramatic and bitter passages of "The Responsibility of Intellectuals," and their detestation was mutual. It is a tribute to both men that they agreed to sign the same document.

42. Hans J. Morgenthau, "Inquisition in Czechoslovakia," *NYRB*, December 4, 1969, pp. 20–21.

43. Diana Trilling and Mary McCarthy, "Ideology and Vietnam" (an exchange of letters), *NYRB*, February 29, 1968, pp. 32–34.

44. Noam Chomsky, "A Reply to Joseph Alsop, *NYRB*, August 21, 1969, p. 38. The citation comes from Chomsky's article "The Menace of Liberal Scholarship," *NYRB* January 2, 1969, p. 37. Chomsky's consistency can be demonstrated elsewhere, even in a militant text like "The Responsibility of Intellectuals" of 1967. In discussing Vietnam war dissent Chomsky notes that some of the most activist and uncompromising protesters were psychologists, chemists, mathematicians, and philosophers, "just as, incidentally, those most vocal in protest in the Soviet Union are generally physicists, literary intellectuals, and other remote from the exercise of power": "The Responsibility of Intellectuals," as reprinted in *American Power and the New Mandarins* (New York: Vintage Books), p. 334. See also Chomsky's 1969 statement that "there may have been a time when American policy in Vietnam was a debatable matter. This time is long past. It is no more debatable than the Italian war in Abyssinia or the Russian suppression of Hungarian freedom": *American Power and the New Mandarins*, p. 9.

45. Anthony Lewis, "Torture in Hanoi," *NYRB*, March 7, 1974, pp. 6–10. In his conclusion Lewis reviews the overwhelming evidence that America bombed civilian targets, destroyed a hospital and a medical school, and did so much more, turning much of the population of Indochina into refugees. "We have poisoned the vegetation and cratered the land, making a desert and calling it peace."

But none of these horrors "can justify ignoring the torture of Americans in North Vietnam or denying that it happened.... Wrongs do not cancel each other out." Sounding very much like Pierre-Henri Simon in *Contre la torture*, Lewis denounces torture as "the ultimate violation of the human spirit." It can never be excused and is "an absolute evil. If we once wink at torture because of our political inclinations—some of us accept it in Chile, say, and others in North Vietnam—there is no stopping."

Cf. also, "Torture in Three Countries," *NYRB*, May 31, 1973, pp. 37–38. The *NYRB* reprinted here the text of statements from an Amnesty International conference. One of the three countries in which the use of torture was documented was the Soviet Union.

46. Karl E. Mayer, "Who's in Charge Here?" *NYRB*, September 10, 1964, p. 4. (A review of David Wise and Thomas B. Ross, *The Invisible Government*)

47. Paul Goodman, "The Liberal Victory," *NYRB*, December 3, 1964, p. 7.

48. Henry Steele Commager, "Common Sense," *NYRB*, December 5, 1968, pp. 3–4. (A review of a collection of Stone's writings on Vietnam, *In a Time of Torment*)

49. I. F. Stone, "The Wrong War," *NYRB*, December 17, 1964, pp. 11, 14.

50. Malcolm W. Browne's *The New Face of War*, and David Halberstam's *The Making of a Quagmire*.

51. I. F. Stone, "Vietnam: An Exercise in Self-Delusion," *NYRB*, April 27, 1965, pp. 4–6. The tone of this article is still dispassionate. Stone does not specifically discuss torture and napalm, but sections from Malcolm Browne's book that do were reprinted adjoining Stone's text by the *NYRB* editors.

52. John K. Fairbank, "How to Deal with the Chinese Revolution, *NYRB*, February 17, 1965, p. 10.

53. I. F. Stone, "McNamara and Tonkin Bay: The Unanswered Questions," *NYRB*, March 28, 1968, pp. 5–12; I. F. Stone, "An Appeal to Averell Harriman," *NYRB*, June 19, 1969, pp. 5–8. This article both offers an excellent example of Stone's powers as an investigative journalist and shows that even at this late date, at a time when many antiwar activists had moved far beyond this approach, Stone sees value in a careful, precisely reasoned appeal to a political figure. Stone analyzed Nixon's peace proposals in several articles, especially "The Hidden Traps in Nixon's Peace Plan," *NYRB*, March 9, 1972, pp. 13–17.

54. I. F. Stone, "Nixon's Blitzkrieg," *NYRB*, January 25, 1973, pp. 13–16.

55. Jean Lacouture, "Vietnam: The Lessons of War," *NYRB*, March 3, 1966, pp. 3–5.

56. I. F. Stone, "Keep 'Em Flying," *NYRB*, January 20, 1966, p. 5.

57. As shown in Elizabeth Hardwick's March 1966 *NYRB* review of Sartre's *The Condemned of Altona*, with its several levels of resonances and direct paralleling with Algeria, already discussed in Chapter 2.

222   Notes

58. As cited in Dwight Macdonald, "A Day at the White House," *NYRB*, July 15, 1965, p. 10. For the full text, see Richard F. Shepard, "Robert Lowell Rebuffs Johnson as Protest over Foreign Policy," *New York Times*, June 3, 1965, pp. 1–2. In the same article Shepherd quotes Lewis Mumford's May 19, 1965, statement denouncing the United States' political and military policy in Vietnam as a "moral outrage" and an "abject failure." (Mumford at the time was president of the American Academy of Arts and Letters)

59. Cited in Richard F. Shepard, "20 Writers and Artists Endorse Poet's Rebuff of the President," *New York Times*, June 4, 1965, p. 2.

60. Macdonald, "A Day at the White House," p. 15. Macdonald decided to attend the ceremonies, though he actively campaigned against the war while on the White House grounds, naturally causing anger and embarrassment on the part of his hosts. It is interesting to compare Macdonald's description of what went on with that of the festival's principal organizer, Professor Eric Goldman, who from 1964 to 1966 served as a special consultant to President Johnson. See Goldman's "The White House and the Intellectuals," *Harper's* 238, no. 1424 (January 1969): 31–46.

61. Irving Howe, Michael Harrington, Bayard Rustin, Lewis Coser, Penn Kimble, "The Vietnam Protest," *NYRB*, November 25, 1965, pp. 12–13. Through 1966, these channels of protest included the teach-ins and the early petition campaign. Kimble was at the time chairman of the New York chapter of the SDS (Students for a Democratic Society, an organization that later became much more radical).

62. Murray Kempton, "Growing Old with the New Left," *NYRB*, January 26, 1967, p. 32.

63. Chomsky had presented a draft as a lecture at Harvard in the spring of 1966, and an early version was published in a small-circulation journal, *Mosaic*, in June of that year. The *NYRB* text, which appeared in the February 23, 1967, issue, was also published in Theodore Roszak, ed., *The Dissenting Academy* (New York: Pantheon, 1968); and in a collection of Chomsky's writings from the Vietnam period, *American Power and the New Mandarins*, pp. 323–66. I shall cite from the latter edition, as it is more readily accessible.

64. Noam Chomsky, "The Responsibility of Intellectuals," as reprinted in *American Power and the New Mandarins*, p. 325.

65. Ibid., p. 333, citing de Gaulle's memoirs. Chomsky goes on to say that de Gaulle's remark needs revision, as America's will to

power is less "cloaked in idealism" than "drowned in fatuity. And academic intellectuals have made their unique contribution to this sorry picture."

66. Ibid., pp. 337–38.

67. Ibid., pp. 352–53.

68. Ibid., p. 353.

69. Ibid., pp. 358–59.

70. See the general discussion of Stage 3 in Chapter 2.

71. Noam Chomsky, "On Resistance," from the *NYRB* of December 7, 1967, as reprinted in *American Power and the New Mandarins*, p. 380.

72. Ibid., p. 385.

73. As William Styron wrote in 1967, after having "signed countless petitions and participated in protests and 'read-ins,' I am beginning to feel that writers and intellectuals are totally impotent in the face of these events." He wonders whether T. S. Eliot was right regarding the Spanish civil war, and (quoting Eliot) "at least a few men of letters should remain isolated" Woolf and Bagguley, eds., *Authors Takes Sides on Vietnam*, p. 72. Cf. also Jerry Rubin, "An Emergency Letter to My Brothers and Sisters in the Movement," *NYRB*, February 13, 1969, pp. 27–29. For further background, see Vogelgesang, *The Long Dark Night of the Soul*, pp. 156–79; and Kadushin, *The American Intellectual Elite*, esp. pp. 355–56, on the alienation and withdrawal of American intellectuals in the early 1970s.

74. Leo Marx, "Letter—The Responsibility of Intellectuals," *NYRB*, November 9, 1967, p. 35.

75. Ibid.

76. Andrew Kopkind, "The New Left: Chicago and After," *NYRB*, September 28, 1967, pp. 3–4.

77. "Noam Chomsky, Chad Walsh, and William X, "An Exchange on Resistance," *NYRB*, February 1, 1968, p. 29. This did begin to change in 1968 and 1969, with the shift at *Time* magazine, for example. See Stanley Karnow, *Vietnam: A History* (New York: Viking, 1983), pp. 488–89. Karnow was the senior correspondent for *Time–Life* in Southeast Asia in the early 1960s.

78. Chomsky et al., "An Exchange on Resistance," p. 30.

79. Noam Chomsky, "Mayday: The Case for Civil Disobedience," *NYRB*, June 17, 1971, p. 24. This was a rough and nasty demonstration, involving by this time significant numbers of extremely angry Vietnam veterans and their guerrilla-theater ap-

proach. Chomsky himself was "a minor—and to be honest, reluctant—participant." He was worried that young people would resort to real violence and that a new contempt for the law was emerging.

80. Martin Bernal, "What Is It About the Vietnamese?" *NYRB*, October 5, 1972, p. 27.

81. Gerald D. Berreman and Frederick Crews, "Vietnam Commencements," *NYRB*, March 28, 1968, p. 38.

82. Among the many examples, see Clarence Brown, "A Commencement Oration," *NYRB*, June 29, 1972, p. 38. Brown, whose name had not appeared in any of the petitions we examined, was a professor of comparative literature at Princeton and a translator of Mandelstam, the great poet who had suffered at the hands of the Stalinist regime—not likely to make him pro-Soviet or its ally. However, Brown was so appalled by what Americans and their sophisticated weaponry had done to the Vietnamese people that he composed a brilliant alternative commencement speech, a striking combination of sick humor and deadly seriousness, which he imagined delivering standing naked, emphasizing the contrast between the stately academic procession and the horrors of Vietnam.

83. "Peace Evening," *NYRB*, April 22, 1971, p. 61; David Marr, "Imprisoned Monks," *NYRB*, March 11, 1971, p. 45; Rennie Davis, Richard A. Falk, and Robert Greenblatt, "The Way to End the War: The Statement of Ngo Cong Duc," *NYRB*, November 5, 1970, pp. 17–22. This is just the smallest sampling. There were humanitarian calls for aid to Vietnamese civilians, a project for a "people's peace treaty," projects to aid American soldiers with legal claims against the army, and much more.

84. Conor Cruise O'Brien, "Confessions of the Last American," *NYRB*, June 20, 1968, pp. 16–18; Paul Lauter and Florence Howe, "The Draft and Its Opposition," *NYRB*, June 20, 1968, pp. 25–31. Cf. also Richard J. Barnet's careful analysis of the negotiations with the North Vietnamese, which were slowly getting under way in the summer of 1968: "The North Vietnamese in Paris," *NYRB*, October 24, 1968, pp. 17–23.

85. Elizabeth Hardwick, "Chicago," *NYRB*, September 26, 1968, p. 5.

86. Margot Hentoff, "Notes from a Plague Year," *NYRB*, November 7, 1968, pp. 23–25.

87. Arthur Waskow, "Opposition," *NYRB*, October 24, 1968, p. 37.

88. Small, *Johnson, Nixon, and the Doves*, esp. p. 193.

89. See, for example, Nina S. Adams, "Man in the Middle," *NYRB*, September 11, 1969, pp. 42–44.

90. Daniel Berrigan, *The Trial of the Catonsville Nine* (Boston: Beacon Press, 1970). The drama was created by combining sections of the trial transcript with various citations, including a passage from Sartre's *The Condemned of Altona*.

91. Francine du Plessix Gray, "The Ultra-Resistance," *NYRB*, September 25, 1969, pp. 11–22. See also Gray's two long and enlightening articles on the 1972 Harrisburg, Pennsylvania, trial of seven radicals (six were Roman Catholics) falsely accused of plotting to blow up the Pentagon: "Harrisburg: The Politics of Salvation," *NYRB*, June 1, 1972, pp. 34–40; and June 15, 1972, pp. 14–21.

92. See Francine du Plessix Gray, "Address to the Democratic Town Committee of Newtown, Conn.," *NYRB*, May 6, 1971, pp. 17–22; and Francine du Plessix Gray, "Danbury Prison Strike," *NYRB*, September 2, 1971, p. 34.

93. Daniel Berrigan, "Letter from the Underground," *NYRB*, August 13, 1970, pp. 34–35; Daniel Berrigan, "On 'The Dark Night of the Soul,'" *NYRB*, October 22, 1970, pp. 10–11.

94. Daniel Berrigan and Robert Coles, "A Dialogue Underground," *NYRB*, March 11, 1971, p. 19.

95. Ibid., p. 22.

96. Daniel Berrigan and Robert Coles, "Dialogue Underground: Inside and Outside the Church," *NYRB*, April 8, 1971, p. 21.

97. "New Mobilization," *NYRB*, November 6, 1969, p. 46; "November Mobilization," *NYRB*, November 20, 1969, pp. 51–52; "Conference on Vietnam," November 20, 1969, p. 54.

98. See Peter Babcox, "The Committee to Defend the Conspiracy," *NYRB*, June 19, 1969, pp. 37–38; Emma Rothchild, "Notes from a Political Trial," *NYRB*, July 10, 1969, pp. 20–26; Herbert L. Packer, "The Conspiracy Weapon," *NYRB*, November 6, 1969, pp. 24–30; Jason Epstein, "The Trial of Bobby Seale," *NYRB*, December 4, 1969, pp. 35–50.

99. Clayton Fritchey, "An Unimpeachable Source Who Can Be Identified," *NYRB*, September 25, 1969, p. 29.

100. On the upsurge of activism after Kent State, see esp. Small, *Johnson, Nixon and the Doves*, pp. 202–5.

101. Paul Goodman, "On the Massacre at Kent State," *NYRB*, June 4, 1970, p. 43. Reprinted with permission from the *New York Review of Books*. Copyright © 1970 Nyrev, Inc.

102. Lawrence Stone, "Princeton in the Nation's Service," *NYRB*, June 18, 1970, p. 7.

103. Ibid., p. 9. Cf. Stanley Diamond and Edward Nell's article on the upheavals at the New School for Social Research. After Cambodia, students at the New School assumed that increased activism was logical, as "they believed the tradition of the New School to be one of engagement and 'relevance'." Instead, there was a bitter confrontation and a three-week sit-in. "The Old School at the New School," *NYRB*, June 18, 1970, p. 38.

104. Jonathan Mirsky, "Nuremberg and Vietnam," *NYRB*, August 12, 1971, p. 30. See also Richard Wasserstrom, "Criminal Behavior," *NYRB*, June 3, 1971, pp. 8–13.

105. Among the many *NYRB* articles on the Pentagon papers, see especially Hannah Arendt, "Lying in Politics: Reflections on the Pentagon Papers," *NYRB*, November 18, 1971, pp. 30–39; Arthur Schlesinger, Jr., "Eyeless in Indochina," *NYRB*, October 21, 1971, pp. 23–32; Noam Chomsky, "Vietnam: How Government Became Wolves," *NYRB*, June 15, 1972, pp. 23–31.

106. I. F. Stone, "The Fakery in Nixon's Peace Offer," *NYRB*, February 24, 1972, p. 15; I. F. Stone, "The Hidden Traps in Nixon's Peace Plan," *NYRB*, March 9, 1972, pp. 13–17. When Stone asked an unnamed government source about our policy toward free elections in South Vietnam, he was told, "The notion of a free election, ... does not come naturally to Leninists." Measured by such a standard, Stone adds, "U.S. policy on elections in South Vietnam can claim to be consistently Leninist." We have demonstrated a curiously schizophrenic capacity "to go on ravaging Indochina in the name of self-determination when on another level of consciousness everyone is aware that there hasn't been a free election in South Vietnam since we took over from the French" (p. 16).

107. Henry Steele Commager, "The Case for Amnesty," *NYRB*, April 6, 1972, pp. 23–25.

108. Frances FitzGerald, "The Offensive: The View from Vietnam," *NYRB*, May 18, 1972, pp. 6–13.

109. I. F. Stone, "Nixon's War Gamble and Why It Won't Work," *NYRB*, June 1, 1972, p. 15.

110. I. F. Stone, "Will the War Go on Until 1976?" *NYRB*, September 21, 1972, pp. 14–16.

111. Basil T. Paquet, "Is Anyone Guilty? If So, Who?" *NYRB*, September 21, 1972, p. 37. Medina's trial was superbly covered by Mary McCarthy, published as a separate book, and included in the collection *The Seventeenth Degree* (New York: Harcourt Brace Jovanovich, 1974).

112. See Commager's exchange of correspondence with Michael

K. Garrity, a draft resister who was at the time in prison, in the *NYRB*, June 15, 1972, pp. 36–38.

113. Henry Steele Commager, "The Defeat of America," *NYRB*, October 5, 1972, p. 7.

114. Ibid., p. 11. "How sobering," Commager added, "that fifteen years before 1984 our government should invent a doublethink as dishonest as that imagined by Orwell." Frances FitzGerald addressed the same theme in the next issue of the *NYRB*. To replace the "nonsense words 'security,' 'pacification,' and 'development' with the real names for U.S. policies would have been to acknowledge the commission of war crimes as defined at Nuremberg" "The Invisible Country," *NYRB*, October 19, 1972, p. 25.

115. Commager, "The Defeat of America," p. 12.

116. As Richard J. Barnet succinctly put it, the task of Vietnamization in Nixon's mind was "to disengage the country [the United States] from the war without losing it": "Nixon's Plan to Save the World," *NYRB*, November 16, 1972, p. 15.

117. Commager, "The Defeat of America," p. 13. With his historian's perspective, Commager understood that Americans had been spoiled by victory and thus found it hard to accept the "elementary lesson of history, that some wars are so deeply immoral that they must be lost."

118. Frances FitzGerald, "Can the War End?" *NYRB*, February 22, 1973, p. 13. Like almost everything else pertaining to Vietnam, the memorial was highly controversial, and a large literature has sprung up concerning it. But its calm, beauty, sensitivity, and evocative power are coming more and more to be recognized, and according to some journalistic accounts, by 1990 it had become Washington's most beloved monument.

119. See I. F. Stone, "Toward a Third Indochina War," *NYRB*, March 8, 1973, pp. 17–20.

120. Cf. William Shawcross, "How Thieu Hangs On," *NYRB*, July 18, 1974, p. 17. Shawcross observed that conservatively estimated, there had been seventy thousand Vietnamese killed between January 1973 and June 1974; that is, in that short period of "peace" more Asians had died than did Americans during our decade of involvement in the war.

121. "Appeal for the Release of Vietnamese Political Prisoners," *NYRB*, April 19, 1973, p. 42. See also Joseph Buttinger, "Thieu's Prisoners," *NYRB*, June 14, 1973, pp. 20–24. Buttinger, who had been writing on Vietnam since 1955, made a careful and conservative estimate that there were more than 200,000 held in the Saigon re-

gime's jails at this time. Arrests increased just before and even after
the cease-fire agreement, and the continued use of torture was well
documented. Despite his years of study of Vietnamese history, Buttinger did not feel that he could predict the eventual outcome of the
struggle for control of the entire country. He was convinced, however, on the basis of many interviews, that unless there were free
political life in the South, which would mean the release of all the
political opponents of Thieu then in prison, "South Vietnam's present government will sooner or later collapse, and the South [will] be
reunited with the North under a Communist-dominated regime"
(p. 24).

122. Karl Bissinger, Joe Chaikin, Noom Chomsky, Martin Du-
berman, Andrea Dworkin, Florence Falk, Richard Falk, Robert Jay
Lifton, Grace Paley, George Wald, "Nobel Peace Prize," *NYRB*, October 4, 1973, p. 38. Seven out of the twelve then current members
of the Department of History at Vassar signed this nominating petition, which was widely circulated. The 1974 Nobel Peace Prize was
awarded jointly to former Prime Minister Eisaku Sato of Japan and
Sean MacBride of Ireland.

123. Francine du Plessix Gray, "Old Times," *NYRB*, February
22, 1973, p. 25. One of the militant members of the VVAW (Vietnam
Veterans Against the War) remarked in 1972: "Our life is being
against the war. When the war ends then *we* end as people." Cited
in J. Glenn Gray, "Back," *NYRB*, June 28, 1973, p. 22.

Recalling his antiwar activism during a 1987 historical conference, Jacques Julliard made the same point in strikingly similar language. He observed that the Algerian War was experienced by
students, especially by Christian students, as a "totalitarian event. I
mean by this an event so entered every facet and every moment of
their lives, that it was for them a foundational event, a generational
fact. . . . I think that there has been nothing similar since. May 1968
was much too brief and in the end was experienced at second hand
[i.e., through radio and television] by the great majority of students.
. . . I remember reflecting with a few militant friends at the UNEF
and the JEC [National Student Union and the Christian youth move-
ment] 'But what are we going to do when the Algerian War has
ended?' " Jacques Julliard, "Débat," in Francois Bédarida and
Etienne Fouilloux, eds., *La Guerre d'Algérie et les chrétiens*, Institut
d'histoire du temps présent, cahier no. 9 (October 1988): 140.

124. Elizabeth Hardwick, "The Meaning of Vietnam," *NYRB*,
June 12, 1975, p. 28. Susan Sontag was glad for the DRV (Democratic
Republic of Vietnam) victory, but not festive or rejoicing. "They"

may have won, but "we" have not. By "we" she meant the antiwar intelligentsia, the base of whose critique was fundamentally moral. "Those of us who raged against this unjust war and its unbearable atrocities reached the limit of our influence when the Sullen Majority turned against the war for quite other reasons—because it was interminable, wasteful, or bungled": Susan Sontag, "The Meaning of Vietnam," *NYRB*, June 12, 1975, p. 24.

125. Christopher Lasch, "The Meaning of Vietnam," *NYRB*, June 12, 1975, p. 28.

126. Norman Mailer, "The Meaning of Vietnam," *NYRB*, June 12, 1975, pp. 26–27.

127. Mary McCarthy, "The Meaning of Vietnam," *NYRB*, June 12, 1975, pp. 25–26.

128. Noam Chomsky, "The Meaning of Vietnam," *NYRB*, June 12, 1975, pp. 31–32.

129. Sheldon Wolin, "The Meaning of Vietnam," *NYRB*, June 12, 1975, p. 23.

130. Gary Wills, "The Meaning of Vietnam," *NYRB*, June 12, 1975, p. 24.

Chapter 5

1. George C. Herring, *America's Longest War: The United States and Vietnam, 1950–1975* (New York: Wiley, 1979), p. 250.

2. See, for example, Isabelle Lambert, "Vingt ans après," in Jean-Pierre Rioux, ed., *La guerre d'Algérie et les Français* (Paris: Fayard, 1990), p. 559.

3. Gloria Emerson, *Winners and Losers: Battles, Retreats, Gains, Losses and Ruins from the Vietnam War* (New York: Harcourt Brace Jovanovich, 1978), p. 37. Emerson goes on to observe that "the quarreling was fierce; sometimes it did not seem as if the war alone could be the reason for the hatred." Emerson cites as one example a box of letters, many of them anti-Semitic, that were sent to the journalist Seymour Hersh, who in 1969 had been the first to expose the Mylai massacre. I would agree with Emerson, with the qualifier that the internal social tensions caused by the Vietnam War permitted varieties of latent racism to come to the surface.

4. Robert J. Lifton, "Victims of Hiroshima," *NYRB*, April 25, 1968, p. 36.

5. Eugene Genovese, "Self-Evident Truths?," *NYRB*, December 19, 1968, p. 36.

6. Michael Harrington, Letter to the Editors, *NYRB*, January 2, 1969, p. 41. Both Harrington and Macdonald were well-known and active opponents of the Vietnam War. The issue they happened to be debating here was the New York City school crisis.

7. André Mandouze, "Débat," in François Bédarida and Etienne Fouilloux, eds, *La Guerre d'Algérie et les chrétiens*, Institut d'histoire du temps présent, cahier no. 9 (October 1988): 150. The French is *nous nous sommes mis à plat*. One wonders whether Mandouze had in mind Herbert Marcuse's conception of the modern individual's being flattened out into one dimension, popularized in his enormously influential *One-Dimensional Man* of 1964, a work that has been translated into French and is well known there.

8. Paul Clay Sorum, *Intellectuals and Decolonization in France* (Chapel Hill: University of North Carolina Press, 1977), p. 244.

9. Sandy Vogelgesang, *The Long Dark Night of the Soul* (New York: Harper & Row, 1974), p. 160.

10. Gilles Lipovetsky, *L'Empire de l'éphémère: La Mode et son destin dans les sociétés modernes* (Paris: Gallimard, 1987). Cf. especially p. 333: "A temporary participation, à la carte, has been substituted for engagement body and soul; one devotes as much time and money as one wishes, one mobilizes himself when he wishes, in the manner that he wishes, conforming to the basic desire for individual autonomy. It is the time of minimal engagement."

11. Roland Jaccard, "Le tragique de la légèreté," *Le Monde* (international ed.), November 11–18, 1987, p. 12.

12. Alain Finkielkraut, "Un militant de l'insignifiance," *Le Monde* (international ed.), November 11–18, 1987, p. 12.

13. Bernard-Henri Lévy, *Éloge des intellectuels* (Paris: Grasset, 1987), pp. 10, 12.

14. Ibid., p. 32.

15. Ibid., p. 37. Lévy does carry his argument further and lay out an interesting set of reasons behind the great popularity of the Sartron and what it stands for. They include the loss of faith in truth, reason, and justice, and the decline of belief in fixed, static, hierarchical values that can be methodically articulated. Also the disappearance of the recognition in the society of *clercs* that there is a kind of dignity in abstract culture, which quoting Sartre, permits the intellectual to leave his ivory tower to get involved in "what is none of his business" (*ce qui ne le regardait pas*) (p. 45). With ever-growing specialization we are witnessing the end of "humanities" and "general culture."

16. Ibid., pp. 126–27. (*démodé* in French)

17. H. Stuart Hughes, *Sophisticated Rebels: The Political Culture of European Dissent, 1968–1987* (Cambridge, Mass.: Harvard University Press, 1988).

18. Pascal Ory and Jean-François Sirinelli, *Les Intellectuels en France, de l'affaire Dreyfus à nos jours* (Paris: Armand Colin, 1986), pp. 47, 113, 189, 242. In an important theoretical article, Sirinelli makes a similar point, arguing that "the curve of their [the intellectuals'] intervention has been globally rising." He believes that there is "a history of the fluctuations of the engagements of the *clercs* to be written": "Une Histoire en chantier: L'Histoire des intellectuels," *Vingtième siècle*, no. 9 (January–March 1986): 108. It is my hope that this book has been a fragment of a response to Sirinelli's call.

19. Ory and Sirinelli, *Les Intellectuels en France*, p. 197. The authors believe that one can legitimately speak of a "generation" of the Algerian War, an age cohort born in the 1930s. Eight separate annual contingents of draftees were sent to Algeria, and the imprint of the war on the intellectual segment of this generation was "crucial" (p. 201).

20. Ibid., p. 223.

21. Ibid., pp. 237–38.

22. Ibid., p. 244. They also observe that it is possible that new theories will appear that claim total explanatory power and will become once again the focus of sharp debate among intellectuals (p. 237). Earlier in their work, when discussing engagement during the Dreyfus affair, they make a very interesting observation that would merit verification through a series of case studies. They suggest that in a period of reflux or withdrawal, (*repli*) artists will reconvert themselves to avant-garde movements, retreating into intellectual creation, "to the detriment of political engagement" (pp. 46–47).

23. It is possible that historians looking back at the Reagan presidency will comment on the brilliant success this militantly unintellectual leader had in suppressing intellectual engagement against his person and his administration. He kept at bay many supposedly shrewd people, who perhaps unwisely scorned him. The informal polls that I have seen, taken at colleges like Vassar, suggest that around 80 percent of the professoriate remained opposed to him and voted against him in 1980 and 1984. It is almost as if Reagan possessed an intuitive sense that told him when to stop to avoid mobilizing the American intelligentsia. One thinks of Grenada, a rapid action lasting just one day, or the caution exercised in Nicaragua.

At least in the first year of his presidency, President George Bush followed his predecessor closely in this regard. The invasion on Panama in December 1989 was over in a matter of hours, and the American casualties were few. One or two teach-ins around the country were organized in response to what to a neutral outside observer would seem like a rather blatant use of military force to eliminate a regime deemed undesirable by Washington. (I know there was a teach-in at the University of Wisconsin.) But Panama was quickly forgotten; it never became a cause for the intelligentsia, and no significant engagement was generated.

24. Mary McCarthy and Diana Trilling, "On Withdrawing from Vietnam: An Exchange," *NYRB*, January 18, 1968, p. 10.

25. Michael Ferber, "On Being Indicted," *NYRB*, April 25, 1968, p. 14. (Italics his)

Epilogue

1. A. L. (André Laude), "Paix séparée," *Esprit* 30, no. 306 (May 1962): 782.

2. This is Robert Frank's argument. He believes that "the identity crisis after the Algerian War was generational and not national." See note 4.

3. Bernard W. Sigg, *Le silence et la honte: Nevroses de la guerre d'Algérie* (Paris: Messidor, 1989). The nearly identical psychiatric problems suffered by American veterans of Vietnam have been the subject of intense study, and a massive literature has resulted. The drug addiction and suicide rates appear higher in this country, whereas the alcoholism rate is probably lower. A good estimate is that by 1986, 80,000 Vietnam veterans had taken their own lives since returning home from the war, considerably more than the 59,000 who were killed in combat. The estimate of a half-million sufferers from PTSD is a conservative one taken from newspaper accounts of interviews with psychiatrists. Stanley Karnow cites a Veterans Administration psychiatrist who estimated that 700,000 Vietnam veterans have or will be afflicted with it. See Stanley Karnow, *Vietnam: A History* (New York: Viking, 1983), p. 25.

4. I have borrowed this phrase from "Troubles de la mémoire française," Robert Frank's extraordinarily evocative and useful essay, which both summarizes the results of his students' researches and develops highly original hypotheses. It was published in Jean-Pierre Rioux, ed., *La Guerre d'Algérie et les Français* (Paris: Fayard, 1990), pp. 603–7.

5. Frank, "Troubles de la mémoire française," p. 603. This observation has been made with extreme frequency and as early as October 1962 by Jean-Marie Domenach (see Chapter 3). Cf. also, Pierre Nora on the "strange silence" that reigns over a "phantom Algeria": "L'Algérie fantôme," *L'histoire*, no. 43 (1982): 9.

6. Claude Liazu, "Le contingent: Entre silence et discours ancien combattant," in Rioux, ed., *La Guerre d'Algérie et les Français*, p. 513.

7. George C. Herring, *America's Longest War: The United States and Vietnam, 1950–1975* (New York: Wiley, 1979), p. 264. In her detailed study of the way that the French press dealt with the Algerian War, Robert Frank's student Isabelle Lambert uses almost identical language to that of George Herring, speaking of "the collective and voluntary amnesia that surrounds the Algerian War": "Vingt ans après," in Rioux, ed., *La Guerre d'Algérie et les Français*, p. 557.

8. As quoted from an interview with the BBC, in Michael Charlton and Anthony Moncrieff, eds., *Many Reasons Why: The American Involvement in Vietnam* (New York: Hill & Wang, 1978), p. 243.

9. Philip Caputo, *A Rumor of War* (New York: Ballantine, 1978), p. 213.

10. Frank, "Troubles de la mémoire française," p. 604. An alternative translation would be "screening," and psychologists speak of "screen memory." Technically the French word *occultation* refers to the action of masking a source of light, as in an eclipse.

11. Ibid., p. 605.

12. Ibid., p. 606.

13. Frédéric Rouyard, "La Bataille du 19 mars," in Rioux, ed., *La Guerre d'Algérie et les Français*, pp. 545–48.

14. Frank, "Troubles de la mémoire française," p. 607.

15. Rouyard, "La Bataille du 19 mars," p. 552.

16. The structure, organization, and purposes of these organizations have been carefully analyzed by Joëlle Hureau in "Associations et souvenirs chez les Français rapatriés d'Algérie," in Rioux, ed., *La Guerre d'Algérie et les Français*, pp. 517–25.

17. As cited in Anne Roche, "La Perte et la parole: Témoignages oraux des Pieds-Noirs, in Rioux, ed., *La Guerre d'Algérie et les Français*, p. 536, interview no. 377. This is part of a massive oral history project, undertaken since 1977, which now has 413 recorded and transcribed interviews.

18. Guy Pervillé, "Les accords d'Évian et les relations franco-algériennes," in Rioux, ed., *La Guerre d'Algérie et les Français*," p. 492.

BIBLIOGRAPHY

The following bibliography is highly selective, designed to direct the interested reader to some of the most useful published sources on the Algerian and Vietnam wars and on the manifestations of intellectual engagement during those conflicts. A few of the most important general works analyzing the social role of intellectuals are also listed.

France and Algeria

Alleg, Henri. *La Question*. Paris: Éditions de Minuit, 1958.

Aron, Raymond. *L'Algérie et la république*. Paris: Plon, 1958.

————. *Mémoires*. Paris: Julliard 1983.

————. *La Tragédie algérienne*. Paris: Plon, 1957.

Aron, Robert, François Lavagne, Janine Feller, Yvette Garnier-Rizet, eds. *Les Origines de la guerre d'Algérie*. Paris: Fayard, 1962.

Azema, Jean-Pierre, Jean-Pierre Rioux, and Henry Rousso. "Les Guerres franco-francaises," *Vingtiéme siécle*, no. 5 (January–March 1985): 3–5.

Bédarida, François, and Etienne Fouilloux, eds. *La Guerre d'Algérie et les chrétiens*. Paris: Cahiers de l'Institut d'histoire du temps présent, no. 9, October 1988.

Broyelle, Claudie, and Jacques Broyelle. *Les Illusions retrouvées: Sartre a toujours raison contre Camus*. Paris: Grasset, 1982.

Bruhat, Jean. *Il n'est jamais trop tard: Souvenirs*. Paris: Albin Michel, 1983.

Camus, Albert. *Actuelles III, Chronique algérienne, 1939–1958*. Paris: Gallimard, 1958.

————. "Les Déclarations de Stockholm," in *Essais*, vol. 2, pp. 1881–82. Paris: Gallimard, Bibliothèque de la Pléiade, 1965.

————. "Discours à l'Académie suédoise," in *Les Prix Nobel en 1957*, pp. 47–50. Stockholm: Imprimerie Royal P. A. Norsted & Söner, 1958.

————. *Le Mythe de Sisyphe*. Paris: Gallimard, 1942.

Cohen-Solal, Annie. *Sartre*. Paris: Gallimard, 1985.

Daniel, Jean. *De Gaulle et l'Algérie*. Paris: Éditions du Seuil, 1986.

Domenach, Jean-Marie. *Ce que je crois*. Paris: Grasset, 1978.

Droz, Bernard. "Le Cas très singulier de la guerre d'Algérie," *Vingtième siècle*, no. 5 (January–March 1985): 81–90.

Droz, Bernard, and Evelyne Lever. *Histoire de la guerre d'Algérie*. Paris: Éditions du Seuil, 1982. Includes a good general bibliography.

Esprit, December 1954 through November 1962, 211 articles. See the notes for Chapter 3 for detailed citations.

Girardet, Raoul. *L'idée coloniale en France de 1871 à 1962*. Rev. ed. Paris: Livre de Poche, 1979.

Hamon, Hervé, and Patrick Rotman. *Les Porteurs de valises. La Résistance française à la guerre d'Algérie*. Paris: Albin Michel, 1979.

Haroun, Ali. *La 7e Wilaya: La Guerre du FLN en France*. Paris:Éditions du Seuil, 1986.

Horne, Alistair. *A Savage War of Peace: Algeria 1954–1962*. New York: Penguin Books, 1979. Includes a bibliography with some English sources.

Hughes, H. Stuart. *The Obstructed Path: French Social Thought in the Years of Desperation, 1930–1960*. New York: Harper & Row, 1968.

———. *Sophisticated Rebels: The Political Culture of European Dissent, 1968–1987*. Cambridge, Mass.: Harvard University Press, 1988.

Lévy, Bernard-Henri. *Éloge des intellectuels*. Paris: Grasset, 1987.

Lipovetsky, Gilles. *L'Empire de l'Éphémère: La Mode et son destin dans les sociétés modernes*. Paris: Gallimard, 1987.

Lottman, Herbert R. *Albert Camus*. Garden City, N.Y.: Doubleday, 1979.

Maschino, Maurice. *L'Engagement*. Paris: Maspero, 1961.

Maspero, François, ed. *Le droit à l'insoumission*. Paris: Maspero, 1961.

Mauriac, François. *Le Nouveau Bloc-Notes, 1958–1960*. Paris: Flammarion, 1961.

McCarthy, Patrick. *Camus*. New York: Random House, 1982.

Le Monde. Paris. France's most respected newspaper provided excellent coverage of the Algerian War, with at least one article almost every day for eight years. The following are a few of the most important articles dealing with the war and the French intelligentsia. When the articles were signed, the authors are indicated.

"Un Appel pour le salut et le renouveau de l'Algérie française," April 21, 1956, p. 5.

"Après une soutenance symbolique à la Sorbonne M. Maurice Audin obtient le doctorat ès sciences,"December 3, 1957, p. 2.

Birmann, Dominique. "Albert Camus a exposé aux étudiants suédois son attitude devant le problème Algérien," December 14, 1957, p. 4.

Cesbron, Gilbert. "Dos à dos," September 29, 1960, p. 1.

"Le Comité Maurice Audin," October 18, 1961, p. 16.

Duverger, Maurice. "Les Deux Trahisons," April 27, 1960, p. 1.

"Les événements d'Afrique du Nord," November 9, 1955, p. 1.

Fauvet, Jacques. "La Gangrène," June 20 1959, p. 1; and "Les Sanctions contre les artistes et fonctionnaires signataires du manifeste sur l'insoumission," September 30, 1960, pp. 1, 6.

Legris, Michel. "*Contre la torture* de P.-H. Simon," March 13, 1957, pp. 1–2.

Marrou, Henri. "France ma patrie," April 5, 1956, p. 2.

"Le Meeting du comité des intellectuels contre la poursuite de la guerre," January 29–30, 1956, p. 3.

"Plusieurs personnalités se groupent au sein d'un comité d'action contre la poursuite de la guerre en Afrique du Nord," November 6–7, 1955, p. 4.

"Des Professeurs à la Sorbonne expriment leur adhésion à la politique gouvernementale, May 23, 1956, p. 3.

"Signé par le Maréchal Juin, un manifeste condamne 'les professeurs de trahison,' " October 13, 1960, p. 1.

"Sommes-nous les 'vaincus de Hitler'?" March 13, 1957.

Suffert, Georges. "Un Récidiviste," December 5, 1956, p. 5.

Vidal-Naquet, Pierre. "La Guerre révolutionnaire et la tragédie des harkis," November 11–12, 1962, p. 11.

Morin, Edgar. *Autocritique.* 3rd. ed. Paris: Éditions du Seuil, 1975.

Mus, Paul. *Guerre sans visage. Lettres commentées du Sous-Lieutenant Émile Mus.* Paris: Éditions du Seuil, 1961.

La Nef. Paris. Especially the issues of April 1957 (Albert Memmi on Albert Camus); February 1958 (Claude Roy on censorship); January 1959 (the Kabyle writer Jean Amrouche on irrational factors as driving on the war); and the double issue of October 1962–January 1963, entirely devoted to the Algerian War, with important articles by Georges Houdin on the Christian response to the war and by Michel Crouzet on the "Battle of the French Intellectuals."

238 *Bibliography*

O'Brien, Conor Cruise. *Albert Camus of Europe and Africa.* New York: Viking, 1970.

Ory, Pascal, and Jean-François Sirinelli. *Les Intellectuels en France, de l'affaire Dreyfus à nos jours.* Paris: Armand Colin, 1986.

Parmelin, Hélène. *Libérez les Communistes!* Paris: Stock, 1979.

Rioux, Jean-Pierre. *La France de la Quatrième République.* Vol. 2, *L'Expansion et l'impuissance.* Paris: Éditions du Seuil, 1983.

————, ed. *La Guerre d'Algérie et les Français.* Paris: Fayard, 1990. The published proceedings of an extremely important international colloquium sponsored by the Conseil National de Recherche Scientifique, held in Paris in December 1988. Includes fifty-nine papers by specialists from seven countries (Algeria among them) and an excellent bibliography up-to-date through 1989. This is an example of the kind of intellectual collaboration that if applied to the United States and Vietnam could produce significant results, helping bring the Vietnam War into the domain of history.

————, ed. *La Guerre d'Algérie et les intellectuels français.* Paris: Cahiers de l'Institut d'histoire du temps présent, no. 10, November 1988. A more specialized collaborative effort. Very important to understanding the role of the French intellectuals during the Algerian War. Published in book form, Brussels: Éditions Complexe, 1991.

Sartre, Jean-Paul. "Préface," to Frantz Fanon, *Les Damnés de la terre*; pp. 9–26. Paris: Maspero, 1961.

————. *Situations,* V. *Colonialisme et néo-colonialisme.* Paris: Gallimard, 1964.

Sigg, Bernard W. *Le Silence et la honte: Nevroses de la guerre d'Algérie.* Paris: Messidor, 1989.

Simon, Pierre-Henri. *Contre la torture.* Paris: Éditions du Seuil, 1957.

Sirinelli, Jean-François, "Une Histoire en chantier: L'Histoire des intellectuels," *Vingtième siècle,* no. 9 (January–March 1986): 97–108.

Smith, Tony. *The French Stake in Algeria, 1945–1962.* Ithaca, N.Y.: Cornell University Press, 1978.

Sorum, Paul Clay. *Intellectuals and Decolonization in France.* Chapel Hill: University of North Carolina Press, 1977. Includes the best available bibliography of materials published through 1976 on the French intellectuals and the Algerian War.

Talbott, John. "The Strange Death of Maurice Audin," *Virginia Quarterly Review* 52, no. 2 (Spring 1976): 224–42.

————. *The War Without a Name: France in Algeria 1954–1962.* New York: Knopf, 1980. An excellent brief introduction. The best place for the nonspecialist to begin. Includes a useful bibliography.

Les Temps modernes, May 1955 through October 1962, 148 articles. See the notes for Chapter 3 for detailed citations.

Tricot, Bernard. *Les Sentiers de la paix, Algérie 1958–1962*. Paris: Plon, 1972.

Verdès-Leroux, Jeannine. *Le reveil des somnambules: Le Parti communiste, les intellectuels, et la culture (1956–1986)*. Paris: Fayard/ Éditions de Minuit, 1987.

———. *Au service du parti: Le Parti communiste, les intellectuels, et la culture (1944–1956)*. Paris: Fayard/Éditions de Minuit, 1983.

Vidal-Naquet, Pierre. "Une Fidélité têtue: La Résistance française à la guerre d'Algérie," *Vingtiàeme siècle*, no. 10 (April–June 1986): 3–18.

———. *La Torture dans la république*. Paris: Éditions de Minuit, 1972.

Winock, Michel. "Les Affaires Dreyfus," *Vingtième siècle*, no. 5 (January–March 1985): 19–37.

———. *La République se meurt: Chroniques 1956–1958*. Paris: Éditions du Seuil, 1978.

America and Vietnam

Armor, David J., Joseph P. Giocquinta, R. Gordon McIntosh, Diana E. H. Russell "Professors' Attitudes Toward the Vietnam War," *Public Opinion Quarterly* 31, no. 2 (Summer 1967): 159–75.

Baritz, Loren. *Backfire: A History of How American Culture Led Us into Vietnam and Made Us Fight the Way We Did*. New York: Morrow, 1985.

Bentley, Eric. "Treason of the Experts," *The Nation*, December 13, 1965, pp. 466–70.

Berrigan, Daniel. *Consequences, Truth, and . . . .* New York: Macmillan, 1971.

———. *The Trial of the Catonsville Nine*. Boston: Beacon Press, 1970.

Berrigan, Daniel, and Lee Lockwood. *Absurd Convictions, Modest Hopes*. New York: Vintage Books, 1973.

Bryan, C. D. B. *Friendly Fire*. New York: Putnam, 1976.

Burns, Richard Dean, and Milton Leitenberg, *The Wars in Vietnam, Cambodia and Laos, 1945–1982: A Bibliographic Guide*. Santa Barbara, Calif.: ABC Clio, 1984. This is a massive effort, superbly organized, including chronologies, maps, and indices to newspapers. For anyone wishing to undertake specialized work on America and Vietnam, this is the place to begin. There are 6,202 citations, although for a bibliography dealing specifically with the intelligentsia during the Vietnam War years this work is less helpful than Vogelgesang. The listings for the entire antiwar movement number only 210. See also Louis A. Peake.

Caputo, Philip. *A Rumor of War.* New York: Ballantine, 1978.

Charlton, Michael, and Anthony Moncrieff. *Many Reasons Why: The American Involvement in Vietnam.* New York: Hill & Wang, 1978.

Chomsky, Noam. *American Power and the New Mandarins.* New York: Vintage Books, 1969.

————. *At War with Asia.* New York: Pantheon, 1970.

Chomsky, Noam, and E. S. Herman. *Bains de sang.* Translated by Marie-Odile Faye. Paris: Seghers/Laffont, 1975. Never published in English.

Collier, Peter, and David Horowitz. *Destructive Generation: Second Thoughts About the Sixties.* New York: Summit Books, 1989.

————, eds. *Second Thoughts: Former Radicals Look Back at the Sixties.* Lanham, Md.: Madison Books, 1989.

DeBenedetti, Charles. *An American Ordeal: The Antiwar Movement of the Vietam Era.* Syracuse, N.Y.: Syracuse University Press, 1990.

Ellsberg, Daniel. *Papers on the War.* New York: Simon & Schuster, 1972, especially the essay dealing with the Nuremberg parallels, "The Responsiblity of Officials in a Criminal War," pp. 275–309.

Emerson, Gloria. *Winners and Losers: Battles, Retreats, Gains, Losses and Ruins from the Vietnam War.* New York: Random House, 1978.

Fall, Bernard B. *Last Reflections on a War.* New York: Doubleday, 1967.

————. *Viet-Nam Witness, 1953–1966.* New York: Praeger, 1966.

FitzGerald, Frances. *Fire in the Lake: The Vietnamese and the Americans in Vietnam.* Boston: Little, Brown, 1972.

Glessing, Robert J. *The Underground Press in America.* Bloomington: Indiana University Press 1970.

Goldman, Eric. "The White House and the Intellectuals," *Harpers* 238, no. 1424 (January 1969): 31–46.

Gray, Francine du Plessix. *Divine Disobedience: Profiles in Catholic Radicalism.* New York: Vintage Books, 1971.

Hellman, John. *American Myth and the Legacy of Vietnam*. New York: Columbia University Press, 1986.

Herr, Michael. *Dispatches*. New York: Knopf, 1977.

Herring, George C. *America's Longest War: The United States and Vietnam, 1950–1975*. New York: Wiley, 1979.

———. "America and Vietnam: The Debate Continues," *American Historical Review* 92, no. 2 (April 1987): 350–62.

———. "Peoples Quite Apart: Americans, South Vietnamese, and the War in Vietnam," *Diplomatic History* 14, no. 1 (Winter 1990): 1–23.

Hoffmann, Stanley. "Vietnam: An Algerian Solution?" *Foreign Policy*, no. 2 (Spring 1971): 3–37.

Howe, Irving. "The New York Intellectuals," *Commentary* 46, no. 4 (October 1968): 29–51.

Jacoby, Russell. *The Last Intellectuals: American Culture in the Age of Academe*. New York: Basic Books, 1987.

Kennedy, Paul. *The Rise and Fall of the Great Powers*. New York: Random House, 1987.

Kadushin, Charles. *The American Intellectual Elite*. Boston: Little, Brown, 1974.

Karnow, Stanley. *Vietnam: A History*. New York: Viking, 1983. A good introduction to the enormous general bibliography on the Vietnam War may be found in Karnow's "Note on Sources," pp. 708–28.

Kolko, Gabriel. *Anatomy of a War: Vietnam, the United States, and the Modern Historical Experience*. New York: Pantheon, 1985.

Ky, Nguyen Cao. *How We Lost the Vietnam War*. New York: Scarborough Books, 1978.

Ladd, Everett Carll. "American University Teachers and Opposition to the Vietnam War," *Minerva* 8 (1970): 541–56.

Lake, Anthony. "The End of an Analogy," *Boston Review* 9, no. 1 (February 1984): 13–15.

Lang, Daniel. *Casualties of War*. New York: McGraw-Hill, 1969.

Lasch, Christopher. "New Curriculum for Teach-Ins," *The Nation*, October 12, 1965, pp. 239–41.

Levy, David W. "The Debate over Vietnam: One Perspective," *Yale Review*, Spring 1974, pp. 335–46.

Mailer, Norman. *The Armies of the Night: History as a Novel. The Novel as History*. New York: New American Library, 1968.

———. *Miami and the Siege of Chicago*. New York: Signet Books, 1968.

———. *Why Are We in Vietnam?* New York: Putnam, 1967.

McCarthy, Mary. *The Seventeenth Degree*. New York: Harcourt Brace

Jovanovich, 1974. A collection of earlier published works on Vietnam, including *Hanoi* and *Medina*. Also "How It Went," a fascinating account of McCarthy's decision to become involved in the antiwar movement.

Mitford, Jessica. *The Trial of Dr. Spock*. New York: Vintage Books, 1970.

Morrow, Lance. "A Bloody Rite of Passage," *Time*, April 15, 1985, pp. 20–31.

Newman, John, with Ann Hilfinger, eds. *Vietnam War Literature: An Annotated Bibliography of Imaginative Works About Americans Fighting in Vietnam*. 2nd. ed. Metuchen, N. J.: Scarecrow Press, 1988.

New York Review of Books, September 1964–June 1975, 262 articles. See the notes for Chapter 4 for detailed citations. Two of the most important post–1975 articles from the *NYRB* are John Gregory Dunne, "The War That Won't Go Away," September 25, 1986; and Jonathan Mirsky, "The War That Will Not End," August 16, 1990.

New York Times. For the nonspecialist, a useful summary of the *Times'* coverage of Vietnam is Arleen Keylin and Suri Boiangui, eds. *Front Page Vietnam*. New York: Arno Press, 1979.

Nobile, Philip. *Intellectual Skywriting: Literary Politics and the New York Review of Books*. New York: Charterhouse, 1974.

Peake, Louis A. *The United States in the Vietnam War, 1954–1975: A Selected, Annotated Bibliography*. New York: Garland Press, 1986. Useful and thorough for coverage of general aspects of the war, with helpful descriptive annotations. The listings total 1,567, with only 20 on the antiwar and peace movement. See also Richard Dean Burns and Milton Leitenberg.

Ramparts, 1962–1975. Especially issues of July 1965, "The Vietnam Lobby"; December 1965, Bernard Fall's "Vietnam Album"; February 1966, "I Quit!" on dissent in the military; April 1966, "The University on the Make," the famous exposé of Michigan State University's role in the policing of South Vietnam; December 1966, the excerpts from Barbara Garson's *MacBird!*; and the double issue of January 1967, the most effective of all of *Ramparts*'s antiwar journalism, with the crucified soldier on the cover and the devastating article "The Children of Vietnam"; and finally, the even more provocative issue of October 1967, with a color photograph on the cover of a handsome young soldier with blood on his hands and groin. (The soldier was actually John Lennon in uniform.)

Rieff, Philip, ed. *On Intellectuals*. Garden City, N.Y.: Doubleday, 1969.

Salisbury, Harrison E. *Behind the Lines—Hanoi*. New York: Harper & Row, 1967.

Schell, Jonathan. *The Village of Ben Suc*. New York: Knopf, 1967.

Schlesinger, Arthur M., Jr. *The Crisis of Confidence: Ideas, Power, and Violence in America*. Boston: Houghton Mifflin, 1969.

Schreiber, E. M. "Opposition to the Vietnam War Among American University Students and Faculty," *British Journal of Sociology* 24, no. 3 (September 1973): 288–302.

Schuman, Howard, and Edward O. Laumann. "Do Most Professors *Support* the War?" *Trans-action* 5, no. 1 (November 1967): 32–35.

Small, Melvin. *Johnson, Nixon, and the Doves*. New Brunswick, N.J.: Rutgers University Press, 1988. The best work on the impact of the antiwar movement on government policy. Superbly researched and carefully argued.

Sontag, Susan. *Trip to Hanoi*. New York: Farrar, Straus & Giroux, 1968.

Taylor, Gordon O. "American Personal Narrative of the War in Vietnam," *American Literature* 52, no. 2 (May 1980): 294–308.

Taylor, Telford. *Nuremberg and Vietnam: An American Tragedy*. New York: Bantam, 1971.

Vogelgesang, Sandy. *The Long Dark Night of the Soul: The American Intellectual Left and the Vietnam War*. New York: Harper & Row, 1974. For further study of the American intellectuals and Vietnam, the best place to begin is the excellent bibliography, up-to-date through 1973. For the period covered, Vogelgesang is more thorough than are the massive general bibliographes assembled by Peake and by Burns and Leitenberg.

Winkler, Karen J. "The Vietnam War Scores Well at the Box Office, but It Fails to Attract Many Researchers," *Chronicle of Higher Education*, September 30, 1987, pp. A4–A6.

Woolf, Cecil, and John Bagguley, eds. *Authors Take Sides on Vietnam*. New York: Simon & Schuster, 1967.

Zaroulis, Nancy, and Gerald Sullivan. *Who Spoke Up? American Protest Against the War in Vietnam, 1963–1975*. Garden City, N.Y.: Doubleday, 1984. A general history of the antiwar movement. Straightforward chronological narrative written in a breezy, journalistic style. Includes a useful bibliography.

Zinn, Howard. *Vietnam: The Logic of Withdrawal*. Boston: Beacon Press, 1967.

INDEX

Abeel, Erica, 42
Abstention, attitude of, 82
Academic faculties, 39, 44. *See also* Petitions; *specific colleges*
Action Committee Against the Pursuit of the War in North Africa, 48–49
Administrative emancipation, 63
Agnew, Spiro T., 130
Ahmia, Cherif, 70
Algerian provisional government in exile (GPRA), 26–27, 34–35
Algerian War. *See also* Cycle of engagement; Parallels between the wars
  casualty count in, 93–94
  chronology for, 14–15
  collective amnesia and, 173–76
  commemoration of, 174–76
  debate over role of intellectuals during, 169
  drugs and, 191n57
  French Catholic hierarchy and, 80–82
  generational impact of, 228n123, 231n19
  as *la guerre de Sartre*, 102–5
  sanitization of, 93–94
  source materials on, 8–9
Algerian workers in France, 34–35, 57, 92–93
Alger républicain (newspaper), 67
Algren, Nelson, 115
Alleg, Henri, 36, 66–67, 100
Alsop, Joseph, 133
Alternative civilian service, 76
Althusser, Louis, 166
Altoona, Pennsylvania, 55–56, 197n44
American antiwar intellectuals
  Camus as model for, 62–71

French intellectuals as model for, 42–47
lack of despair among, 146–47
as majority of American intellectuals, 112–13
"Manifesto of the 121" and, 44–45, 62
studies of, 43–44
women and, 217n22
American intellectuals, attitudes toward Vietnam War among, 113–14, 115–16. *See also* American antiwar intellectuals; Progovernment intellectuals
American periodicals, and opposition to Vietnam War, 129
American society
  civil impact of war in, 120, 156–57, 159–60, 162–64, 197n43
  crisis of identity in, 10
  cultural responsibility and, 37
  paradoxes of Vietnam War and, 147–48
American Special Forces in Vietnam, 17
Amherst College, 158
Amnesty for draft resisters
  in France, 24, 100, 177–78, 211n126
  in U.S., 24, 154, 178
Armies of the Night, The (Mailer), 59
Aron, Raymond, 20, 166–68
Aron, Robert, 20, 100, 185n13
Artistic community, 40
Atlantic Monthly, 129
Auden, W. H., 116
Audin, Maurice, 36, 68–69
Authors Take Sides on Vietnam (Woolf and Bagguley), 115–16

245

from Vietnam, 21, 115–16,
136–37, 197*n*42
Nazi war crimes. *See* Nuremberg
precedents
Negative inspiration, 77
"New Call to Resist Illegitimate
Authority, A" (petition), 124
Newman, John, 9
"New philosophers" group, 166.
See also Finkielkraut, Alain;
Lévy, Bernard-Henri
New Politics Convention (Chicago, 1967), 146
New Republic, 119, 123–24, 129,
219*n*34
New School for Social Research,
225*n*103
New York City peace march
(April, 1967), 128
New Yorker, 129
New York Review of Books
aftermath of Algerian War and,
111
aftermath of Vietnam War and,
157–59
anti-Soviet writings in, 132–33
antiwar petitions and, 119–20,
124, 130
categories of materials in, 133–
34
Chomsky's "Responsibility of
Intellectuals" article and,
141–44
compared with *Esprit*, 76–77
contributors to, 130–31
cycle of engagement in, 133–60
"Meaning of Vietnam, The" article, 159–60
moral consistency of coverage
in, 133
parallels between the wars and,
16–18
as source, 9, 129–31
Stone's writings in, 135–38
varieties of engagement and,
150–55
New York Times, antiwar petitions
in, 117–19, 120–21, 216*n*14
Nixon, Richard, 138, 157, 158
election of, 21, 150
negotiated withdrawal and, 25–
26, 28, 124, 152

petition campaign and, 129
Nizan, Paul, 82
Nobel Peace Prize of 1974, 158,
228*n*122
North Vietnamese, use of torture
by, 133
Novick, Peter, 6
*Nuremberg and Vietnam: An Ameri-
can Tragedy* (Taylor), 153–54
Nuremberg precedents
American antiwar intellectuals
and, 115, 122, 123, 144, 153–
54
counterlegal engagement and,
50–52
Esprit and, 84
I. F. Stone's commentary on,
155–57
parallels between the wars and,
17–18

OAS. *See* Organisation Armée
Secrète
"Obstructed path" metaphor,
208*n*97, 214*n*146
"Open Letter on Vietnam" (peti-
tion), 121
"Open Letter to President John-
son on Vietnam, An" (peti-
tion), 117–19
Oral history, 6–8
Organisation Armée Secrète
(OAS), 24, 75, 84, 92, 203*n*51
Ory, Pascal, 116, 169–70
Ouzegane, Amar, 62

Pacification programs, 31–32,
120, 184*n*4, 212*n*133
Paquet, Basil, 155
Parallels between the wars
Chomsky's essay on responsibility and, 145
in conduct of war, 16–33
documentation of, 130
in intellectual engagement generated by, 42–47
legislative votes of support and,
15–16
Stone's comments on, 137–38
Paras, 184*n*3
Paris de Bollardière, General
Jacques, 184*n*6
Paris-Match, 93–94